# 2nd European NDT & CM Days 2021



Prague, Czech Republic
4 – 7 October 2021

ISBN: 978-1-7138-4140-1

**Printed from e-media with permission by:**

Curran Associates, Inc.
57 Morehouse Lane
Red Hook, NY 12571

**Some format issues inherent in the e-media version may also appear in this print version.**



Printed with permission by Curran Associates, Inc. (2022)

For permission requests, please contact
Brno University of Technology & Czech Society for Non-Destructive Testing
at the address below.

Brno University of Technology
Techniká 2 CZ 616 69 Brno
Czech Republic

Phone: +420 541 147 664

www.cndt.cz

**Additional copies of this publication are available from:**

Curran Associates, Inc.
57 Morehouse Lane
Red Hook, NY 12571 USA
Phone: 845-758-0400
Fax: 845-758-2634
Email: curran@proceedings.com
Web: www.proceedings.com

# TABLE OF CONTENTS

USE OF THE MMM METHOD FOR DETECTION OF CRACKS IN THE WELDED JOINT OF PIPELINES ..... 1
*Jaroslav Brom, Jan Patera, Václav Svoboda*

COMBINING NDTS WITH DIGITAL PHOTOGRAMMETRY SURVEY AND AMBIENT VIBRATION DE-TECTION FOR DIAGNOSIS OF PATHOLOGIES IN HISTORIC ARCHITECTURES TOWARD CONSERVATIVE AND COMPATIBLE MEASURES ..... 10
*Mariella De Fino, Alessandra Fiore, Fabio Fatiguso*

VIBRATIONAL QUALITY CLASSIFICATION OF METALLIC TURBINE BLADES UNDER MEASUREMENT UNCERTAINTY ..... 22
*L. Cheng, V. Yaghoubi, W. Paegpegem, M. Kersemans*

ACOUSTIC EMISSION BASED DAMAGE MONITORING TO STUDY FRACTURE IN CEMENTITIOUS MORTARS UNDER FATIGUE LOADING ..... 29
*Nuhamin E. Deresse, Charlotte Van Steen, Mina Sarem, Stijn François, Els Verstrynge*

DEVELOPMENT OF AN ULTRASONIC TESTING SYSTEM FOR PRESS CONNECTORS OF RAILWAY CATENARY WIRES ..... 39
*Thomas Herzog, Frank Schubert, Stephan Heilmann, Jürgen Michauk, Susan Walter, Henning Heuer*

CONDITION MONITORING OF TOROIDAL ROLLER BEARINGS USING ACOUSTIC EMISSION TECHNIQUE ..... 49
*Housam Mohammad, Frantisek Vlasic, Pavel Mazal*

VIBRATION-BASED FEATURE EXTRACTION FOR ARTIFICIAL DAMAGES IN SMALL SCALE CUTTING DISCS OF TUNNEL BORING MACHINES ..... 58
*Sebastian Priebe, Lukas Brackmann, Arne Rottger, Gunter Meschke, Inka Mueller, Inka Buethe*

NON DESTRUCTIVE EXAMINATION OF CARGO VAPORIZERS AND CONDENSERS IN SHIPS CARRYING LIQUEFIED PETROLEUM GASES IN BULK (LPG CARRIERS)..... 68
*T. Ingale*

BASIC STUDY ON IMPACT ELASTIC WAVE METHOD TO EXPAND APPLICATION SCOPE OF GROUTING CONDITION EVALUATION IN PC TENDON DUCTS ..... 78
*Kentaro Yamashita, Shinichi Hattori, Koki Terasawa, Toshiro Kamada*

AUTOMATED DEFECT DETECTION IN CFRP USING FLASH INFRARED THERMOGRAPHY AND DEEP LEARNING METHOD..... 88
*Zongfei Tong, Saeid Hedayatrasa, Liangliang Cheng, Shejuan Xie, Mathias Kersemans*

ACTIVE THERMOGRAPHY TO DETECT DEFECTS IN WOOD SAMPLES OF INTEREST FOR CULTURAL HERITAGE..... 97
*Massimo Rippa, Pasquale Mormile*

DIGITALIZATION FOR RAILWAY-NDT ..... 102
*Daniel Werner, Andreas Franzen, Frank Henrix, Uwe Philipps, Paul Buschke*

NEXT GENERATION TANK FLOOR SCANNING – TOTAL INSPECTION ..... 111
*Stuart G. Kenny, Matthew Boat*

ACOUSTIC ANISOTROPY IN PRODUCTS OBTAINED BY THE METHOD OF SELECTIVE LASER MELTING ........ 117
*Vitaly V. Muravev, Olga V. Muraveva*

50 YEARS OF CONFERENCES DEFEKTOSKOPIE OR (UN) TOLERABLE LIGHTNESS OF NON-DESTRUCTIVE TESTING IN CZECHOSLOVAKIA ........ 123
*B. Kopec*

VYUŽITÍ DRONÙ PØI DIAGNOSTICKÝCH PRÙZKUMECH STAVEBNÍCH OBJEKTÙ ........ 139
*Michal Jankù, Jiøí Grošek, Josef Stryk*

MOŽNOSTI NEDESTRUKTIVNÍHO ZKOUŠENÍ KØEMÍKOVÝCH MATERIÁLÙ ........ 145
*Pavel Mareš, Jana Veselá*

ZVYŠOVÁNÍ A OVÌØOVÁNÍ ODBORNÉ ZPÙSOBILOSTI PRACOVNÍKÙ NDT PROVÁDÌJÍCÍCH ZKOUŠKY NA VYBRANÝCH ZAØÍZENÍCH JADERNÝCH ELEKTRÁREN. ........ 153
*Josef Neugebauer, Petr Fousek*

POTENTIAL OF ACTIVE THERMOGRAPHY AS A NON-DESTRUCTIVE TESTING METHOD FOR QUALITY ASSURANCE OF WELDED JOINTS ........ 159
*Eugen Prints, Igor Kryukov, Malte Mund, Christian Šrajbr, Klaus Dilger, Stefan Böhm*

NEDESTRUKTIVNÍ TESTOVÁNÍ HLAV VÁLCÙ, SEDEL A VENTILÙ VE VÝROBÌ A SERVISU ........ 171
*Bøetislav Skrbek, Jakub Mráz*

HIGH TEMPERATURE ULTRASONIC TESTING BY PHASED ARRAY TECHNIQUE ........ 179
*J. Vesela, P. Mares*

**Author Index**

# Use of the MMM method for detection of cracks in the welded joint of pipelines

Jaroslav BROM [1], Jan PATERA [1], Václav SVOBODA [2]

[1] Diagnostic and Qualification; Research Centre Rez; Hlavni 130, Rez, 25068 Husinec, Czech Republic;
Phone: +420602286054; e-mail: jaroslav.brom @cvrez.cz;
Phone: +420776460049; e-mail: jan.patera@cvrez.cz
[2] Preditest s.r.o.; Pod visnovkou 1662/23, 140 00 Praha 4, Czech Republic;
Phone: +420602331005; e-mail: svoboda@preditest.cz

**Abstract**
The article describes the process and main outputs from the verification of the MMM (Metal Magnetic Memory) method when used to detect cracks in welded joints in pipelines. The verification procedure was divided into several parts. Prior to the production of test specimens with as-real cracks, MMM measurements were performed on semi-finished test specimens containing crack-free welded joints. This was followed by verification of the MMM method for crack detection without and with thermal insulation and verification of the effect of magnetization on the results of the MMM method.
**Keywords:** Metal magnetic memory, pipeline, welded joint, crack, detection

## 1. Introduction

Research Center Rez (RCR), in cooperation with Preditest, solved a R&D project focused on the use of MMM for crack detection in welded joints within the project of the National Centre for Energy, sub-project DP6 - Development of diagnostic methods for characterization of key components of energy units. The R&D project was divided into several parts, where in the first part before the production of test specimens with as-real crack, the measurement was performed by the MMM method on semi-finished test specimens containing crack-free welded joints. After the production of test specimens with as-real crack, the detection and determination of the dimensions of as-real crack was performed by other NDT methods. This was followed by research aimed at verifying the possibilities of the MMM method for detecting cracks without and with thermal insulation and verifying the effect of magnetization on the results of the MMM method.
The materials for the article are drawn from [1].

## 2. Verification of the use of the MMM method for the detection of cracks in welded joints

### *2.1 Brief description of the MMM method*

The MMM method is an NDT method based on sensing the intensity of the magnetic field Hp in the form of residual magnetization on the surface of the material using one or more probes, each of which senses the magnetic field in two mutually perpendicular planes. Residual magnetization of the material is affected by the production process, machining, cutting, welding, bending, heat treatment, cooling, operational stress, etc. Similar magnetization of the material occurs during the process of material degradation in pipeline operation due to degradation conditions - material fatigue, possible corrosion, etc. the device on which the probes are located is traversed over the material to be measured and the connected encoder also measures the distance over which the inspection is carried out. The signal from the individual channels, together with the distance, is recorded in digital form in the device's memory and displayed on the display in graphical form either directly as Hp or in the form of a gradient dHp / dx depending on the distance travelled Lx. The scanning device contains a few measuring probes, with each probe measuring the intensity of the magnetic field in two channels - in the direction of movement of the scanning device the magnetic field intensity is measured by channel x, in the direction perpendicular to the movement of the scanning device the magnetic field intensity is measured by channel y).

Testing by the MMM method was performed within this project using the following measuring technique:

- Tester of Stress Concentration - TSC-7M-16, No П129
- Scanning equipment (three types):
  - Type 1-8 with 4 probes (total 8 channels)
  - Type 2M-475 sensor with 2 probes (total 4 channels)
  - Type 6-8-036 sensor with 8 probes (total 16 channels)
- - SW MMM 4 - system - evaluation software

All measurements were performed at a room temperature of about 20 ° C in order to eliminate the effect of temperature and ensure mutual comparability of all measured results by MMM method.

Figure 1. Example of MMM measuring technique. Evaluation unit and scanning device 1-8M

### *2.2 Test specimens*

The dimensions and materials of the test specimens were chosen based on an agreement with the project partner ČEZ, a.s. such that they correspond to the nozzle No. N6 of the Temelín NPP steam generator, which has a high probability of cracking in the dissimilar metal weld (DMW) of pipes made of 22K (carbon steel) and 08CH18N10T (stainless steel) and with pipe dimensions of 89 x 6 (mm x mm). The basic material of test specimens for the R & D project consisted of equivalent steels 12022.1 (for carbon steel 22K) and X6CrNiTi18-10 (for stainless steel 08CH18N10T). Test specimens contained homogeneous welds made according to WPS used in equipment maintenance.

After the completion of the development of the production of fatigue cracks by high-cyclic loading in steel half-tubes, the production of test specimens was started. The production of cracks first consisted in the production of an artificial defect by the EDM method on half-tube semi-finished test

specimens. EDM notches were used to initiate as-real fatigue cracks by high-cyclic loading. A total of 6 test specimens with as-real cracks were produced – see figure 2.
After completion of the high-cycle loading, a penetration test was performed on the test specimens to confirm or refute the presence of the crack.
After performing the penetration test, an ultrasonic test (PAUT technique) was performed to determine the depth of the defect.

Figure 2. Manufactured test specimens – total 6 pcs of half-tubes made of materials 12022.1 (3 pcs) and X6CrNiTi18-10 (3 pcs).

### *2.3 Program of verification of MMM method for detection of cracks in welded joints*

The verification program was performed in the range of:

1. Measurement of MMM on semi-finished test specimens with crack-free welded joint:
   - Weld joint detection
   - Detection of a welded joint through the applied thermal insulation
2. Measurement of MMM on 6 pieces of test specimens with a crack in homogeneous weld:
   - Detection of as-real defects
   - Detection of as-real defects through the applied thermal insulation
3. Measurement of MMM on 1 piece of test specimen without and with load at the crack
4. Measurements to verify the effect of magnetization and demagnetization on the measurement results by MMM method. These measurements were performed on 2 test specimens (1 carbon steel and 1 austenitic steel) in the range of:
   - Without thermal insulation fitted
   - When thermal insulation is fitted.
5. Verification of the MMM method

All measurements ad 1 to 5 were performed by placing the probe directly on the surface of the test specimen or on the surface of the sheet metal covering the thermal insulation applied to the test specimen.

The direction of movement of the probe was:
Longitudinal (longitudinal across the welded joint, i.e. in the axis of the pipe/half-pipe)
Perimeter (around the circumference of the welded joint, i.e. perpendicular to the axis of the pipe / half-pipe).
The following sections demonstrate the test results.

### *2.3.1 Measurements on semi-finished test specimens with crack-free welded joint*

Summary of measurements:
Measurements of welded joints showed their detectability when scanning in the longitudinal direction. The welds are identifiable in the dHp/dx gradient graphs as bands with higher values. For the detection of welded joints, the results are more pronounced for the gradients in the y-channels, i.e. in the direction perpendicular to the movement of the scanning device.
The measured gradients in ferritic welded joints, in contrast to austenitic welds, show a wider range due to the more extensive thermal influence of the entire weld area – see figure 3.
Based on the measurements, criteria for crack detection in homogeneous welds were proposed for the materials used.
Measurements through thermal insulation, approx. 10 cm, proved the identification of the welded joint and its location along the insulated pipe section, both for ferritic and austenitic welded joints. When applying the MMM method over double insulation, approx. 20 cm, the location of the welded joint was not unambiguously identified.

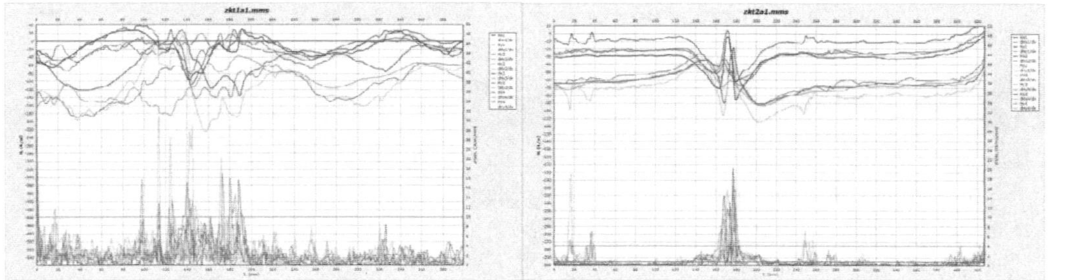

Figure 3: Example of a magnetogram of a ferritic test specimen (left) and an austenitic test specimen (right) with a crack-free weld

### *2.3.2 Measurements on test specimens with a crack in a homogeneous weld*

Measurements of crack welded joints showed the detectability of cracks when scanning in the longitudinal direction according to the criteria proposed for the maximum gradient dHp / dx in Chapter 5.1. The S / N ratio criteria for ferritic steels were not met. Crack weld joints are identifiable in the dHp / dy gradient graphs as bands with higher values. Because half-tube test specimens were used for manufacturing reasons, crack localization was not performed around the circumference of the welded joint.

The values of the dHp / dy gradients were evaluated only for lane 2, which was located above the half-tube axis (the values of the dHp / dx gradients when scanning at the extreme tracks could be affected by the half-tube section).

### *2.3.4 MMM measurement on 1 piece of test specimen without and with load at the crack*

Summary of this measurement

In contrast to the circumferential scan, a longitudinal scan from the outer surface found a significant difference in the measured values and magnetic field strengths Hx, Hy and their gradients between the sensor S1 (above the crack) and S2 (outside the crack) – see figure 4. Significant changes were recorded only on sensor S1. This indicates that while maintaining the distance between sensors S1, S2, the entire length of the defect was not covered when measuring from the outer surface and sensor S2 was most likely outside the defect when passing through the weld.

This result would then confirm that during the longitudinal scan it is possible to detect the presence of a circumferential defect in the welded joint and to distinguish its indication from the indication of the welded joint itself.

Figure 4: Scanning direction and position of the sensors of the two-channel probe when opening the crack defect

### *2.3.5 Measurement on test specimens to verify the effect of magnetization and demagnetization*

Summary of this measurement:

The effect of sample magnetization is manifested by increasing the level of the SCZ gradient at the site of the existing defect by about 2 times. At the same time, changes in the course of the magnetograms appear at the points of application of both poles of the magnetization yoke to the sample. The area of both poles of the magnet is visible. After the demagnetization, the level of the SCZ gradient at the defect site decreases about 2x. Figure 5 shows the magnetograms before and after magnetization.

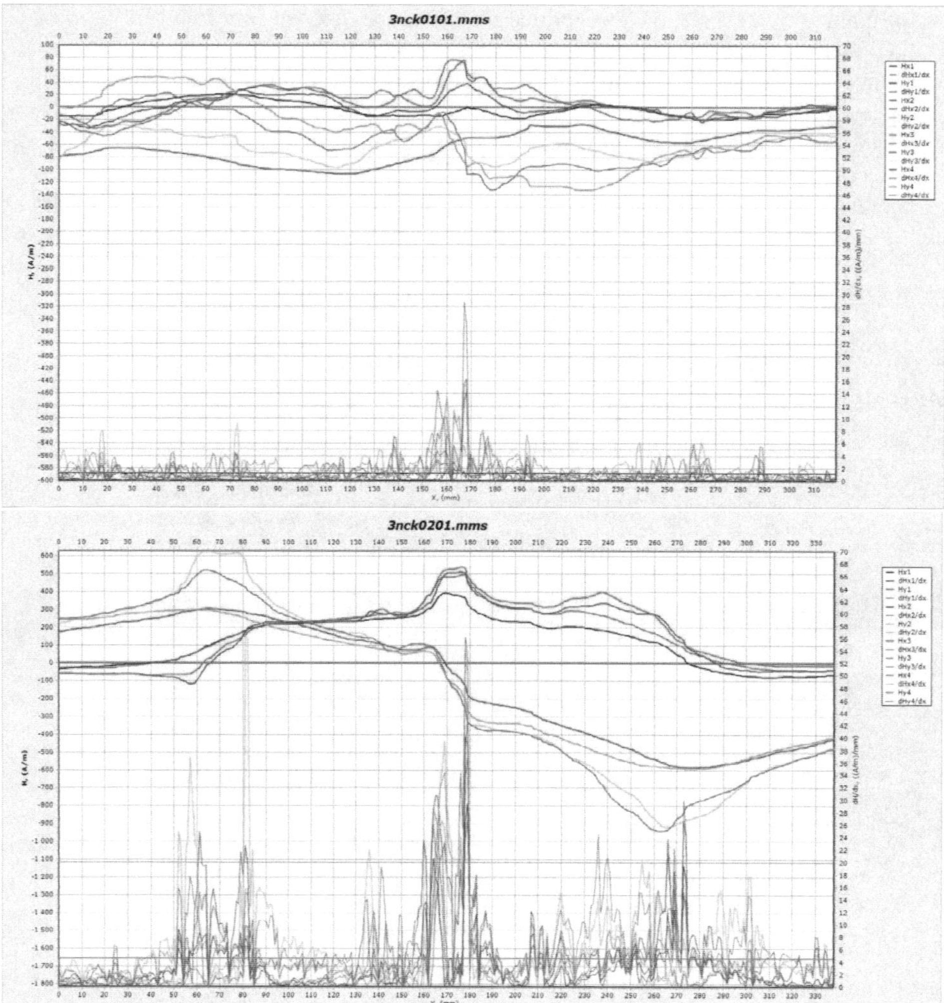


Figure 5: Magnetograms of a ferrite test specimen with a crack on weld with. Up before magnetization, down after magnetization.

### *2.3.6 MMM method verification for detection of cracks in welded joints*

The verification of the MMM method took place on the test specimen of the R&D partner of the ČEZ project. This test specimen was cut from a part of the super-emergency supply pipeline of the steam generator at the Dukovany NPP. The test specimen contains a DMW with a real crack.
Measuring range:
Longitudinal measurement on a test specimen on the marked nine adjacent tracks 1, 2, to 9, parallel to the axis of the pipe perpendicular to the welded joint. These tracks covered the entire circumference of the pipe – see figure 6.
Criteria for MMM technology / method verification:
C1 Identification of the weld joint in all measured tracks
C2 Crack detection based on the existence of a track with a higher dHp/dx gradient
C3 Crack localization agrees with localization according to PAUT technique

Figure 6: Raster with marked paths on the ZT and display of the scanned path

#### *2.3.6.1 Weld joint identification*

The welded joint was identified by the MMM method in all measured tracks.

#### *2.3.6.2 Crack detection*

For data from the 2M-475 type scanner, a crack was identified at track No. 1, where the largest maximum value of the dHp/dx gradient was recorded (83 versus 47 ÷ 74 (A/m)/mm).

#### *2.3.6.3 Crack localization by PAUT technique*

The PAUT measurement showed a crack in the heterogeneous weld near track No. 1.

Based on the above, all criteria C1, C2 and C3 were met.

## 3. Conclusions

1. The R&D program focused on the use of the MMM method for the detection of crack-type defects in welded joints has shown that the MMM method detects the area of the welded joint during longitudinal scanning, the crack location is identifiable by an increase in magnetic field intensity gradient.
2. Measurements around the circumference of the welded joint did not yield usable information.
3. For the testing of welded joints by the MMM method by longitudinal scanning, an inspection procedure may be established, which should include the following principles:
   I. Measurement procedure:
      a. Enter in the form the identification of the inspected object, the date of measurement and the place of measurement, the name of the employee performing the measurement, the degree of certification, the status of the measured places (visual assessment)
      b. Clean inspection areas for coarse dirt and dust
      c. Create a scanning raster for longitudinal measurement so that the paths cover the entire circumference of the tested pipe. The path width is selected according to the width of the selected scanner type. To locate the crack, the tracks should partially overlap. Mark the raster unambiguously.
      d. Take a photo of the raster, including a landmark (place on the object).
      e. If possible, take a macro photograph of the weld.
      f. Check instrument calibration.
      g. Perform your own measurements. Scan each path on the raster at least twice, save it to memory, and write the name of the data file to the form at the same time. The saved files will be used for evaluation after transfer to the PC.
      h. Carry out a preliminary check of the level of gradients on the measured section after the end of the measurement.

   II. Criteria for detection and localization of cracks around the circumference of the welded joint can be designed. Use absolute values of gradients in the "y" direction for evaluation, S / N ratio can be used as an additional evaluation.
      a. The values for the criteria for crack detection by longitudinal scanning are best generated empirically in a similar manner to that used in this project. For empirical determination of the criteria, it is recommended to use the same type of scanning device for measurements without and with a crack.
         In case no data from pilot measurements are available, for the ferritic steel 12022.1 and austenitic steel X6CrNiTi18-10 in the first approximation the values found in this project can be selected:
         - Crack detection criterion for ferritic steel values of the maximum gradient dHp/dx greater than 40 (A/m)/mm
         - Detection criterion for austenitic steel: maximum gradient values dHp/dx greater than 25 (A/m)/mm
      b. Criterion for locating the crack around the circumference of the welded joint by longitudinal scanning
         - location of the measured path with a maximum gradient dHp/dx compared to other measured paths.

4. Measurements through the applied thermal insulation are only usable to determine the location of the weld joint. It cannot be used for crack detection. However, it is recommended to perform pilot measurements on a specific pipe and verify the possibilities of the MMM method through thermal insulation.
5. Magnetization of the welded joint influences the MMM method in the amplification of the measured parameters (acts as a magnifying glass). This can be used in operation to refine the measurement outputs of the MMM method (e.g. more accurate crack location). However, it is recommended to perform a pilot measurement on a specific pipe and to verify the possibilities of the MMM method with magnetization of the welded joint.
6. Verification of the technology was successful on a body with a real crack in a heterogeneous welded joint.
7. When using the MMM method on a newly measured welded joint, it is recommended to perform a pilot measurement on a crack-free welded joint and subsequently to trend the maximum values of dHpy/dy gradients in the measuring paths.
8. Summary of MMM method verification
   The verification of the technology took place on one piece of test specimen with a real crack in a DMW without the possibility of pilot measurement on similar test specimen without/with a crack.
   The MMM method indirectly detects cracks in welded joints. It should always be noted that the MMM method is an NDT method based on sensing the intensity of a magnetic field Hp in the form of residual magnetization on the surface of a material using one or more probes, each of which senses a magnetic field in two planes perpendicular to each other. The crack is manifested mainly by a change in the dHp/dx gradient in the direction perpendicular to the direction of movement (i.e. in the "y" channel of the measuring probes).
   Crack detection and localization can be evaluated by measuring the maximum values of dHpy/dy gradients in the measuring tracks. The evaluation of the signal-to-noise ratio at the cracked location is not very different from the non-cracked locations, so this evaluation can be used as a supplement to the evaluation by the maximum values of the dHpy / dy gradients.
   Verification has shown that MMM methods can be used as non-destructive methods for evaluating the quality of welded joints. For a specific test specimen with a heterogeneous welded joint, a crack detection and localization in the welded joint was performed.

**Acknowledgements**
The presented work has been realized within Institutional Support by Ministry of Industry and Trade of the Czech Republic.

**References**
References should be written in the order in which they appear in the text in the following format:
1. Jan Patera, Jaroslav Brom, ' NCK - Output V1.14.4 Proven technology and summary report for MMM', Technical report No. PL3684, RCR, December 2020.

# Combining NDTs with digital photogrammetry survey and ambient vibration detection for diagnosis of pathologies in historic architectures toward conservative and compatible measures

Mariella DE FINO[1], Alessandra FIORE [2], Fabio FATIGUSO [1]

[1] Department of Civil, Environmental, Land, Construction and Chemistry, Polytechnic of Bari, Bari, Italy
e-mail: mariella.defino@poliba.it, e-mail: fabio.fatiguso@poliba.it
[2] Department of Civil Engineering Sciences and Architecture, Polytechnic of Bari, Bari, Italy,
e-mail: alessandra.fiore@poliba.it

## Abstract

Within the performance assessment of historic architectures, inspection, diagnostic and monitoring methods play a crucial role toward the comprehensive knowledge of both the effects deriving from potential pathologies and the factors affecting the actual behaviour of the system, with specific attention to local vulnerabilities and failure mechanisms. For the purpose, NDTs need to follow an integrated approach, where multidisciplinary methods, including laboratory testing, reality-based survey and detection of static and dynamic parameters, are applied. The paper is going to discuss the above-mentioned issues and present the case study of the Town Hall of Altamura (Ba), in South Italy, where a variety of correlated experimental techniques was carried out in order to assess causes and magnitude of a relevant cracking pattern, related to the facade rotation. Particularly, along with radar scanning, video-endoscopy and direct sampling, the investigation relied on some innovative technologies, including digital photogrammetry for survey of unknown underground structures and detection of ambient vibrations whose analysis allowed assessing not only the global structural safety but also the comfort level.

**Keywords:** Historic architecture, masonry structures, onsite diagnostics; radar scanning; digital photogrammetry; dynamic monitoring

## 1. Introduction

Within the performance assessment of historic architectures, inspection, diagnostics and monitoring methods play a distinctive role compared with most scientific and technical fields, taking into account that the targets are very heterogeneous in terms of characteristics, alterations and surrounding conditions, the investigation practice involves different specialisms and expertise and the preservation of the inherent formal and structural configuration is a key requirement. In fact, they should help address conservative measures meeting high effectiveness and low intrusiveness, by providing comprehensive knowledge and understanding of multi-faceted aspects, including the *effects* deriving from potential pathologies, on the one hand, and the *factors* affecting the actual behaviour of the system, on the other hand, with specific attention to local vulnerabilities and failure mechanisms that are generally difficult to be assessed by analytical procedures.

Particularly, concerning the *effects*, namely cracks, deformations, humidity patterns and surface alterations, experimental measurements might be useful for assessing their typology, extent and progression beyond visual survey and, thus, supporting the sound identification of the occurring anomalies, taking account the bi-univocal relationship between effects and pathologies. Furthermore, concerning the *factors*, investigation tools might be crucial in relating the pathologies to the possible causes, which should be found among a great variety of inner – geometry, materials, construction techniques, load conditions, previous reinforcement works - and outer factors – foundation soil, microclimate, vibrations, surrounding structures. Consequently, the most common onsite NDTs for masonry structures – e.g. thermography, sonic and ultrasonic tests, radar scanning, video-endoscopy - need to follow an integrated approach, where different techniques and methods, including laboratory testing, reality-based survey and detection of static and dynamic parameters, contribute to the identification of the spatial and functional relationships between the building, on the one hand, and its environment, its parts and its transformations, on the other hand.

## 2. Methods and techniques

The scientific community commonly acknowledges NDTs as valuable tools for damage identification and health monitoring of historic buildings [1], by evaluation of texture, morphology and cracking of masonry structures [2], assessment of mechanical properties of stone components [3] [4], identification of failure mechanisms of vulnerable systems, such as slender structures [5]. In all cases, it is underlined the need to correlate extended, qualitative and indirect data from NDTs with limited, quantitative and direct data from DTs, as well as to integrate the onsite diagnostic information with preliminary historical research and subsequent numerical simulation. In fact, documentary, experimental and analytical records are complementary aspects of the diagnostic process, leading to the diagnosis and the restoration, which should meet the well-known criteria of minimal intervention, compatibility and reversibility.

Furthermore, the state-of the-art reveals how some emerging innovative technologies, mainly applied in other fields, such as sciences of representation and mechanical engineering, have been used recently for the architectural heritage, too. Among them, it is worth mention for the present study digital photogrammetry for three-dimensional modelling, as well as analysis of vibrations through dynamic monitoring, since they are both methods supporting high effectiveness and low intrusiveness.

In detail, due to the availability of very affordable hardware and software tools, digital photogrammetry currently offers a reliable, low-budget and user-friendly way to obtain the speedy and accurate 3D reconstruction of an artefact from large datasets of overlapping 2D pictures. Particularly, some automated procedures, i.e. SfM (Structure for Motion) and MVR (Multiview Stereo Reconstruction), enable the identification of matching features in multiple images, initial estimates of camera positions and point coordinates, which are then refined iteratively. The result is a coloured point cloud, namely a set of points associated to x-y-z coordinates and RGB values that provide a photorealistic visualization of the detected artefact. Further computer-based algorithms are able to convert a point cloud into 3D vector data, digital ortho-images and digital surface models. In the field of architectural heritage, there are countless and valuable applications of photogrammetry-based products for representation, diagnosis and management [6] [7] [8] [9]. Particularly, the main advantages of digital photogrammetry are: (i) data acquisition and elaboration by low-cost tools and automated routines that make it suitable even for non-expert users; (ii) operational procedures featured by speedy activities, as well as by small sized and manageable equipment, that might be applied in narrow, difficult to access and fragile contexts; (iii) capability to describe the colour and texture characteristics of the architectural surfaces leading to realistic visual documentation.

Similarly, the experimental assessment through dynamic monitoring has proved to be a reliable way for enhancing the security of heritage buildings and monuments. The analysis of the recorded dynamic response under ambient or forced vibrations is often adopted for: (i) experimental-based analysis of the global dynamic behaviour (ii) evaluation of the structural integrity (iii) calibration of numerical models. Many applications of structural dynamic monitoring techniques to the analysis of historical constructions can be found in the scientific literature [10] [11] [12]. In particular, the protection of historical monuments from road and subway traffic vibrations is a key issue for many city administrations. The structural health of historical constructions is in fact influenced by the effects not only of long-term vibrations induced by vehicles and subways but also of short-term vibrations due for example to construction operations. Such vibrations, even if are not immediately perceivable by people, usually act continuously, so causing a sort of slowly accumulating damage in structures. Small-amplitude vibrations, in general, do not represent an immediate hazard, but over the years they can increase the structural vulnerability of damaged or deteriorated elements [13]. For example, vibrations induced by vehicles and trains may become dangerous especially for structures that were exposed to severe dynamic loads, such as earthquakes, in the past. Similarly, repeated vibrations acting on old masonry structures may lead to deterioration of mortar and then to its strength reduction, with consequent detachment of bricks/blocks. In addition, the excessive roughness of road pavements due to poor maintenance is a possible cause of dynamic stress in adjacent buildings. Therefore, vibrations and dynamic stresses applied to heritage buildings are becoming a growing concern for the institutions managing historical buildings. These issues require an exhaustive effort to assess the severity of ambient vibrations and their effects on heritage buildings.

Within this framework, the herein reported research is going to present and discuss the informative case study of the Town Hall of Altamura (Ba), in South Italy, where a variety of correlated experimental techniques - i.e. radar scanning, video-endoscopy, direct sampling, digital photogrammetry, measurements of displacements and accelerations – was carried out with the purpose of assessing causes and magnitude of a relevant cracking pattern and support decision-making on compatible actions to be undertaken. Particularly, the key role of digital photogrammetry and vibrations analysis as innovative diagnosis tools is discussed.

## 2. Case study

The Town Hall of Altamura stands on the layout of the fifteenth century Convent of San Francesco, substantially transformed in the eighteenth century after the suppression of the monastic orders and then almost totally demolished at the end of the nineteenth century for the construction of spaces and structures intended to house the current function. The latter process was conducted over fifty years, minimally adapting and incorporating the previous constructions. In fact, only at the beginning of the twentieth century, the building assumes its present definitive configuration, with the main front facing Piazza Municipio, the quadrangular plan with a central courtyard and the development on three floor levels, with some underground storage and cistern rooms. The elevation rooms are covered by barrel, pavilion and sail vaults, while the pitched roof is made with a recently built wooden system. All the vertical and horizontal structures are covered with plaster and the wall texture in natural stone is visible only in some points. From the conservative point of view, the north-west corner between Piazza Municipio and Via Teatro Vecchio is of particular interest for this study. In fact, it shows widespread cracking patterns on the facade walls, on the intrados of the ground floor and first floor vaults and on the architraves of the openings, such as to suggest the rotation of the corner portion (Figures 1 and 2). It is interesting to underline the presence of said crack patterns already in a survey carried out in the 90s. In order to plan the investigation campaign, based on the available geometrical survey and the thematic mapping of the visible decay patterns, it was speculated which information were already available on *factors/causes* and *effects* of such a façade rotation and how further assessment by onsite inspection, diagnostic and monitoring techniques was needed to achieve comprehensive knowledge and understanding of the occurring pathology.

Concerning the *inner factors/causes, geometry, materials and construction techniques* were taken into account. Firstly, it was observed that the *geometry* of the building in the area of interest is quite regular, with corresponding walls on all the floors. Nevertheless, it was found that all the vaults are quite thin (≈ 30cm at the ground floor and ≈15cm at the first and second floors) compared to the dimensions of the rooms (up to 6m per side). Moreover, whenever smaller, they transfer the loads on very thin internal walls (≈20cm). The above-mentioned peculiarities might be connected with an unbalanced drift of the vaulted structures on the piers, typically leading to rotation of the perimeter walls. Since the configuration of the foundation level was unknown, the *search and survey of hypogeal structures/spaces* was scheduled with double purpose: to detect geometrical irregularities above the ground level, such as elevation walls not supported at the foot by corresponding foundation walls, as well as to map potential decay, humidity and/or crack patterns of the foundation structures that can be related to the façade rotation as a consequence of foundation anomalous movements. Finally, concerning the technical characteristics of the building components, it was considered that general data from historical research and direct survey was quite poor. Thus, the *assessment of materials and construction techniques* was scheduled, with specific focus on masonry texture, transversal stratigraphy and properties of stone materials of both walls and vaults.

Concerning the *outer factors/causes, foundation soil and ambient vibrations* were examined. However, considering that previous extensive geotechnical tests had ruled out that the bedrock foundation soil was undergoing subsidence or settlement, it was considered that the accelerations due to the traffic, especially during the loading/unloading of goods by big trucks for the nearby local market, can somehow affect the health conditions of the structure. Thus, the *analysis of vibrations* was planned.

Concerning the *effects*, it was observed that the crack patterns were satisfactory assessed in terms of *typology* and *extent* by visual survey. Their progression was assumed quite low, considering the presence of very similar damage evidence even in the survey carried out thirty years before. Nevertheless, the *static monitoring of crack evolution* was scheduled to make sure that some progression was currently occurring. Moreover, the above-mentioned survey of hypogeal structures/spaces was assumed helpful also in view of assessing potential cracks at the foundation level.

Thus, the following activities were carried out, as described in 2.1: *search and survey of underground structures, assessment of materials and construction characteristics, analysis of vibrations and static monitoring of crack evolution.*

Figure 1. From the left: Crack pattern on the facade on Via Teatro Vecchio, on the ground and on the first floor

2.1 Inspection, diagnostics and monitoring

The *search and survey of potential underground rooms* was carried out according to a minimally invasive approach, proceeding with the systematic scanning of all the walking surfaces of the ground floor using 600 MHz frequency radar instrumentation (IDS DAD FastWave single-channel 16-bit radar system and IDS TR600V antenna), compatible with depth of investigation of the order of 1m for the specific components. From an operational point of view, two profiles orthogonal to each other were acquired for each room. Therefore, on the basis of the results of the indirect tests, under the rooms G1 and G3 (Fig.1) it was possible to hypothesize the presence of two underground spaces/structures of which the reflections induced by the intrados surfaces of the covering vaults were visible, differently for the signal absorption of the foundation soil filling (Fig.2). For the sake of shortness, we refer below to the hypogeal space under the room G3.

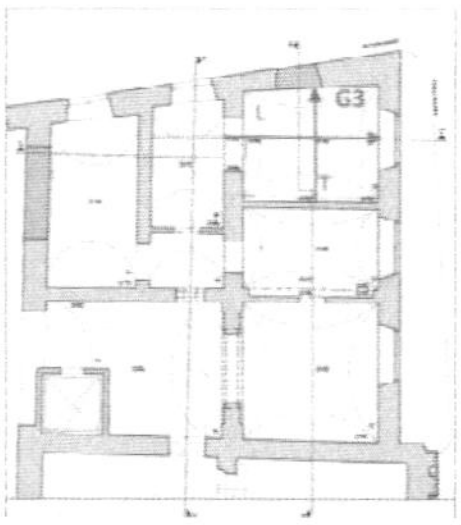


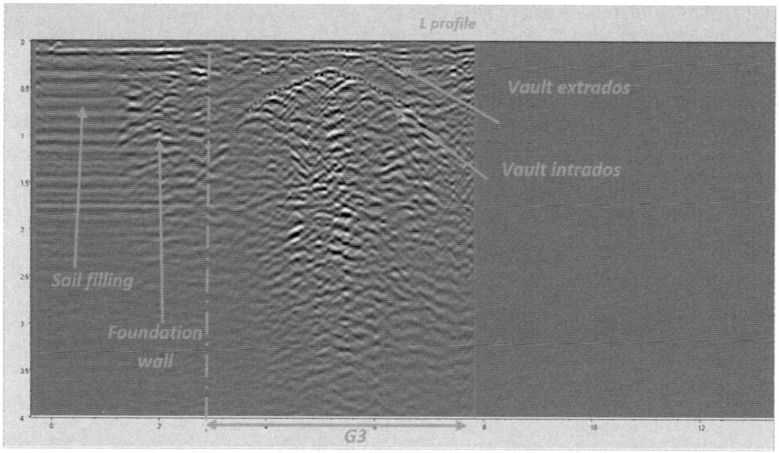


Figure 2. Results from radar scanning in room G3

In order to confirm the hypotheses, 4mm perforations on the floors and video-endoscopies (Ex-tech video-endoscope with HDV-5CAM-3FM probe and HDV600 recording unit) were then performed for a first approach inspection of the underground spaces/structures. The inspection also made it possible to identify the position of some existing trapdoors (Fig. 3).

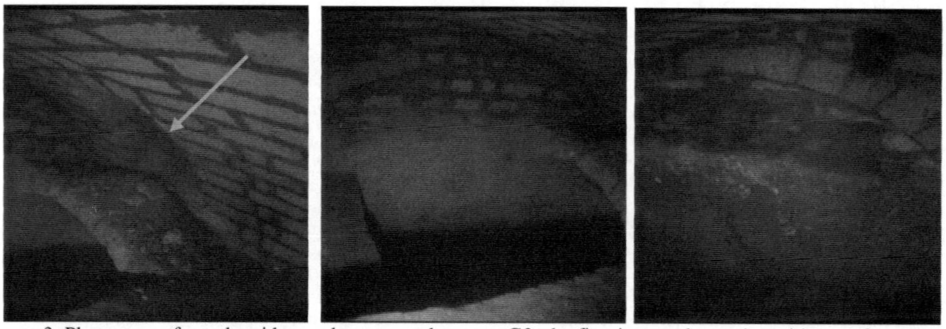

Figure 3. Photograms from the video-endoscopy under room G3: the first image shows the evidence of a trapdoor

Therefore, in order to carry out the geometric and material-constructive survey of the volumes in a quick but accurate way, also with a view to limiting as much as possible the interference with the functions within the premises of the Town Hall, we proceeded to the opening of the existing trapdoors and the direct access for conducting a digital photogrammetric survey, carried out for each room over two hours by a single operator. Specifically, the pictures were acquired by moving the camera (Canon EOS M3 with 24.2 Megapixel resolution) along the perimeter of the room, at different heights and different angles, also with the aid of a telescopic rod, in order to guarantee an overlap between consecutive images equal to 80%. The images were then processed by means of Agisoft Photoscan® software for conversion into a coloured point cloud (Fig.4). The onsite acquisition of some direct measurements made it possible to obtain a three-dimensional scaled model, for the subsequent extraction of representative horizontal and vertical sections (Fig. 5).

The photogrammetric survey of the environment below the room G3 allowed verifying and detailing the observations partially made during the video-endoscopic inspection. The underground volume develops on a rectangular plan and is delimited by two longitudinal walls parallel to the front on Piazza Municipio and two transverse walls parallel to the front on Via Teatro Vecchio. The room has a barrel-vaulted roof, set on the longitudinal walls, and is divided by a series of structural elements. In fact, one observes a masonry wall, with a height equal to the springer of the vault, which crosses it transversely and divides it into two areas. The innermost area has in turn a portion of a barrel vault and an arch, both at heights lower than the springer of the main vault. It should be noted that the internal transverse masonry wall is not aligned with the overlying separation wall between rooms G3 and G5 (Fig.1). From a material-constructive point of view, the vault is made of limestone blocks, arranged in rows along the directives, with over-square mortar joints. This configuration also concerns the walls above the springer. Below, the walls are covered with plaster. The overall state of conservation is good and no cracks were visible on the surfaces.

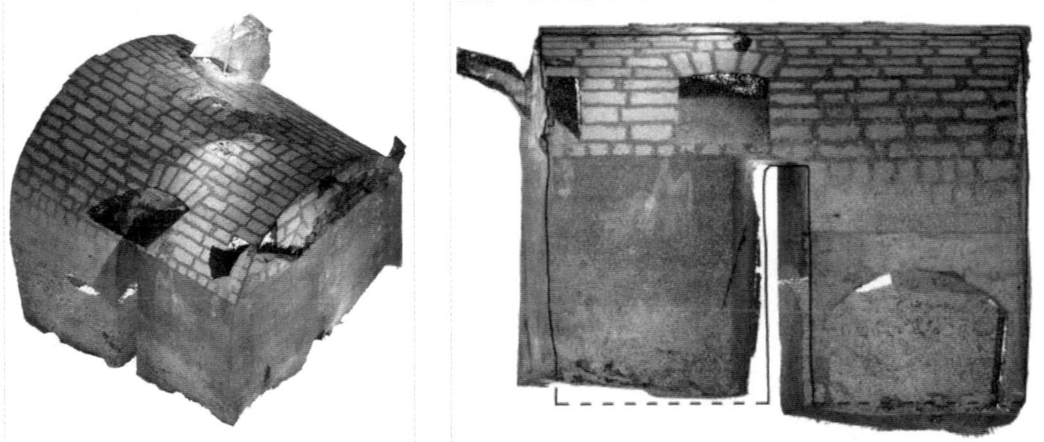

Figure 4. View of the coloured point cloud and ortho-image of the volume under the room G3

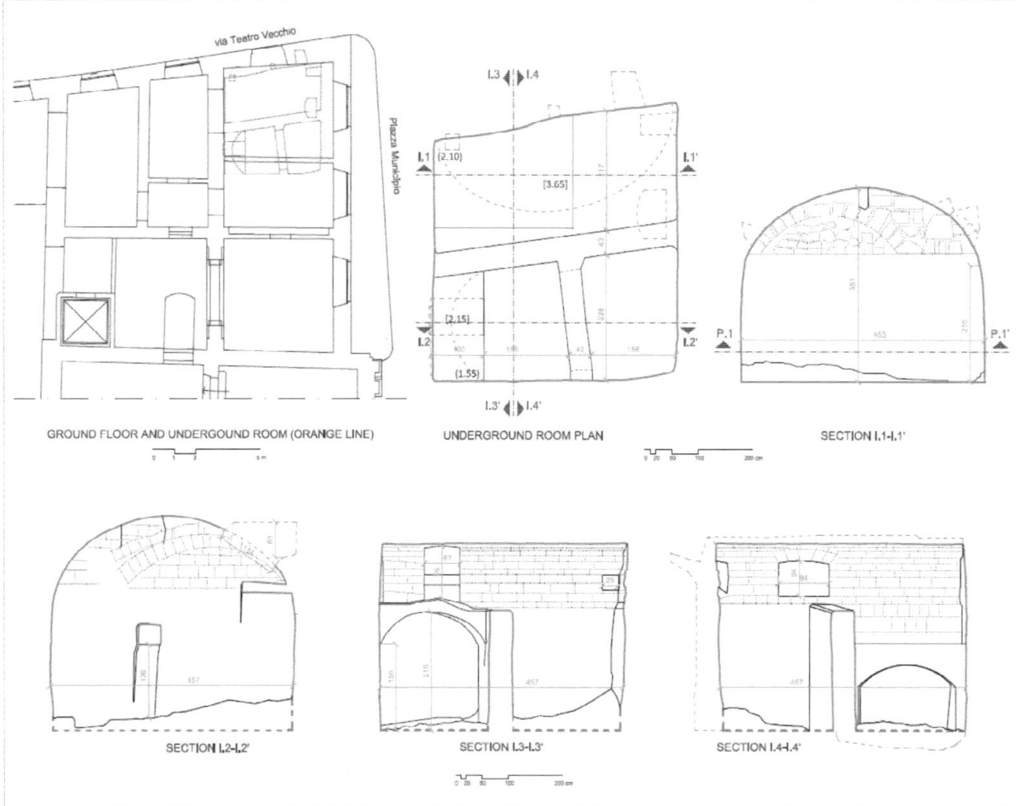


Figure 5. Plan sections of the volume under the room G3 from the coloured point cloud.

The *assessment of materials and construction characteristics* involved a series of direct observations of the façade textures and of the transversal stratigraphies of the walls, both through samples, performed after removing the coating plasters, and by extraction of 80mm cylindrical cores (HUSQVERNS DMS 240 electric core drill with 82mm diameter diamond crown) and subsequent video endoscopies (Ex-tech video-endoscope with HDV-5CAM-3FM probe and HDV600 recording unit). Specifically, it was found the presence at all floor levels of large-sized limestone blocks - 50x25x20cm - with staggered vertical joints, thin mortar edges - about 1cm - and good connections at the corner intersections (Fig. 6). In the cross section, the masonry is multi-layered, with external facing of the type described above and internal core - about 50cm thick - in conglomerate of mortar and stone with good characteristics of homogeneity and compactness. The blocks, later tested in a laboratory, showed an average compressive strength of 3.9 N/mm$^2$. As for the vaults, they have a composite structure, being made of limestone blocks at the springers, above which there are three rows of sheet brick bricks (Fig. 7). Likewise, the filling on the extrados results in a conglomerate of mortar and stone with good characteristics of homogeneity and compactness.

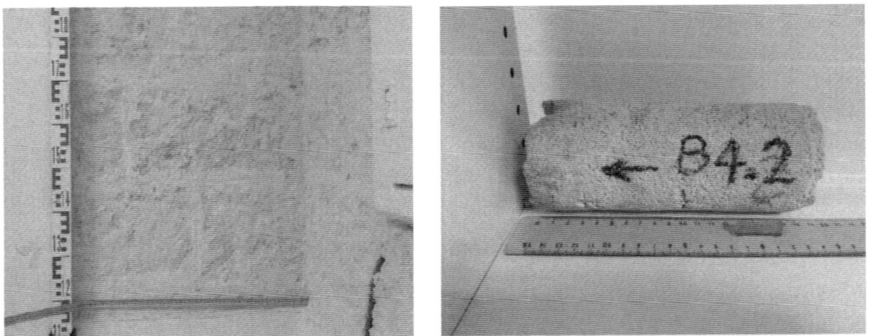


Figure 6. Results from direct sample (left) and borehole drilling (right) of masonry walls

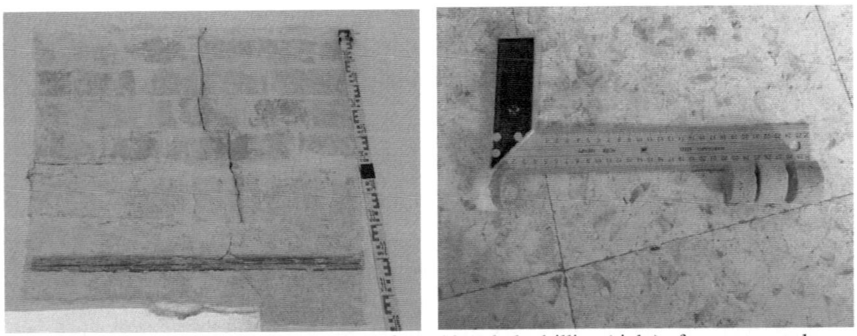

Figure 7. Results from direct sample (left) and borehole drilling (right) of masonry vaults

Concerning the *analysis of the vibrations*, dynamic tests were performed under ambient excitations: traffic was used as the source of excitation. The building was instrumented with six uniaxial piezoelectric accelerometers. Their main properties are: sensitivity 10000 mV/$g$ ($g$ being the gravity acceleration); measurement range ± 0.5 $g$; bandwidth from 0.15 to 1000 Hz; operating temperature from −45 to +82 °C. The accelerometers were installed three on the ground floor and three on the second floor, in the $x$ (2), $y$ (1) and $z$ (3) directions, according to the setup shown in Fig. 8. The dynamic response of the building was recorded using a cyclic buffer with pre-defined length (20 minutes) during the hours mostly exposed to traffic. The vibration frequencies were evaluated separately from the detected accelerations, for each sensor position. For the sake of conciseness, only a few of time and frequency domain results are presented in the following. More precisely, the maximum recorded responses are provided, for each direction, in Fig. 9, and refer to the time intervals 4:50-5:10 a.m. and 6:50-7:10 a.m. For each sensor, the following graphs are reported: *i*) the accelerogram; *ii*) the histogram of accelerations, representing, for each value of acceleration (in abscissa), the corresponding number of times (in ordinate) in which it occurs during the same record; *iii*) the Fast Fourier Transform (FFT). The FFTs of records surely give an idea of the values of the natural frequencies of the Town Hall, through their peak values. By overlapping, for each direction, the obtained FFTs, the natural frequencies result equal to 4.7 Hz in the $x$ (2) direction and 4.5 Hz in the $y$ (1) direction. The main goal of the analysis was to assess the structural integrity of the Town Hall. At this aim, an extensive bibliographic research was carried out, in order to identify recommended upper limits of vibration to which historical constructions and ancient monuments should be subjected. On the basis of it, the thresholds proposed in the document NRC/CNR TT-1900 [14] from the "Canada Institute for Scientific and Technical Information" were selected.

In these guidelines, the following upper limits, in terms of accelerations, are suggested:

- 0.20 m/s² during construction operations,
- 0.02 m/s² for continuous loading (traffic or metro lines)
- 0.05 m/s² for occasional loading.

In the case under examination, the condition concerning continuous loading was considered, that is the threshold 0.02 m/s², valid for all directions $x$, $y$, $z$. In Fig. 9, the above bound equal to 0.02 m/s², corresponding to 0.002 g, is depicted on each acceleration time history. By comparing accelerations with the upper limit, it can be noted that the threshold is never exceeded.

In a second step, it was verified if the vibrations of the Town Hall were acceptable in terms of human perceptibility. In fact human beings are strongly sensitive to vibrations and often suffer from oscillation intensities well below those necessary to overstress structural elements. Therefore, human response to vibrations caused by dynamic loading, such as the passage of vehicles, should be treated as a serviceability limit state and so suitable checked in a verification process. On this topic, the code ISO 2631-2: 1989 [15] provides human perceptibility thresholds for residential and office places under transition vibration in terms of mean Root Mean Square (RMS) of acceleration in m/s². The acceptable values are:

- 0.0036 RMS (m/s²) in the horizontal directions $x$ and $y$;
- 0.005 RMS (m/s²) in the vertical direction $z$.

The mean RMS of accelerations can be calculated as:

$$RMS = \sqrt{\frac{1}{N}\sum_{n=1}^{N}\left|x_n\right|^2} \qquad (1)$$

where $N$ represents the number of samples constituting the record $x(t)$ and $x_n$ is the $n^{th}$ sample. The code ISO 2631-2: 1989 [11] also furnishes multiplying coefficients >1 to be applied to the limit thresholds in function of the building destination use. For the analysed case-study, for each acceleration record, the corresponding mean RMS with the respective threshold is reported in Table 1. It can be argued that the mean RMS values never exceed the comfort thresholds.

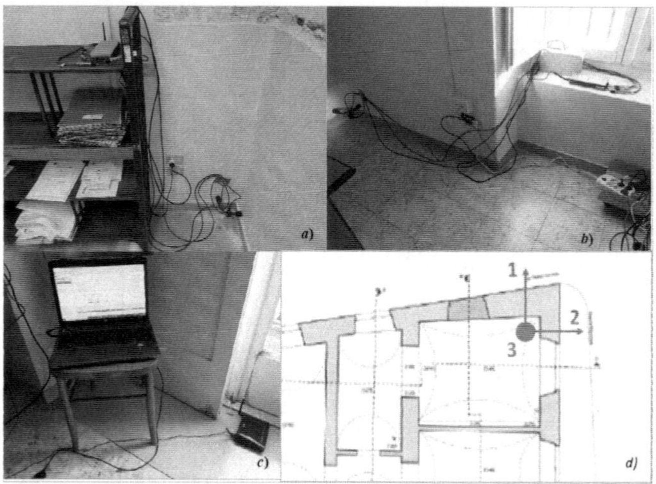


Figure 8. Installation on the ground floor (a) and second floor (b); c) acquiring system; d) directions of accelerometers.

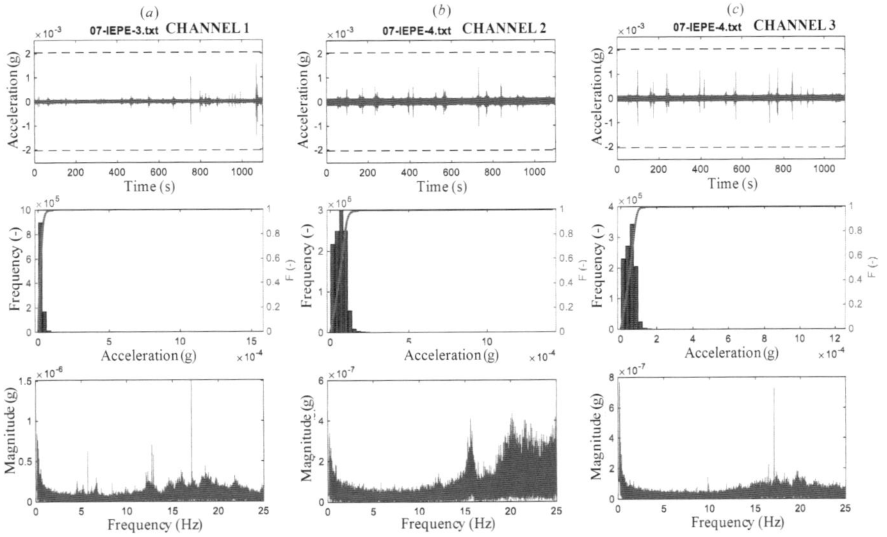


Figure 9: For each channel (in vertical): accelerogram, histogram of accelerations, FFT; a) channel 1, second floor, time interval 4.50-5:10 a.m., date 7th March 2020; b, c) channels 2, 3 respectively, ground floor, time interval 6.50-7:10 a.m., date 7th March 2020.

| RMS in m/s² | | | |
|---|---|---|---|
| **RECORD** | **CHANNEL 1 threshold: 0.0036** | **CHANNEL 2 threshold: 0.0036** | **CHANNEL 3 threshold: 0.005** |
| 06-IEPE | 5.9418e-04 | 7.0990e-04 | 5.9369e-04 |
| 06-IEPE-1 | 4.3619e-04 | 4.6445e-04 | 3.2669e-04 |
| 07-IEPE-2 | 5.4166e-04 | 6.8734e-04 | 5.6028e-04 |
| 07-IEPE-3 | 2.5006e-04 | 2.5236e-04 | 2.0358e-04 |
| 07-IEPE-4 | 5.9103e-04 | 7.1726e-04 | 5.7905e-04 |
| 07-IEPE-5 | 3.3455e-04 | 3.6023e-04 | 2.6186e-04 |
| 07-IEPE-6 | 5.5348e-04 | 7.0657e-04 | 5.6703e-04 |
| 07-IEPE-7 | 2.6875e-04 | 2.5579e-04 | 2.0238e-04 |
| 07-IEPE-8 | 5.5288e-04 | 7.0359e-04 | 5.6936e-04 |
| 07-IEPE-9 | 3.0744e-04 | 3.0214e-04 | 2.3267e-04 |

Table 1: RMS of accelerations. Comparison with comfort thresholds.

Finally, concerning the static *monitoring of crack evolution*, five crackmeters consisting of two stackable polycarbonate plates (FES-TT1 PLUS AVONGARD) were installed across the most significant cracks, with monthly manual readings. For all the measurement bases, after one year, no appreciable vertical/horizontal displacements were detected.

2.2 Discussion of results

The investigation campaign conducted by the integrated use of non-destructive techniques, localized sampling and innovative methods of digital survey and vibration analysis, supported the reliable diagnosis of the occurring pathologies. In particular, the absence of cracking phenomena in the underground rooms was noted, which makes it possible to exclude problems of roto-translation of the basement structures, possibly transmitted to those in elevation. Similarly, the impact of vibrations on the integrity and safety of the building appears to be of little significance. Likewise, the vertical walls appear of excellent workmanship both for the characteristics of resistance of the materials, and for the regularity of the textures and compactness of the internal cores. On the other hand, the criticality of the vaults is confirmed, which, in addition to the reduced thickness, are made of brick blocks of small size, with a large volumetric incidence of the mortar joints. Moreover, in one of the cases in which the vaults are supported by an internal masonry wall of reduced thickness (separation walls G3-G5 in Fig. 1), the wall is in turn supported on the roof vault of an underground space. Overall, therefore, the geometric and material-constructive configuration of the vaulted structures reasonably suggests that there is an unbalanced distribution of the actions that weigh on the facades, which, in the absence of the counterthrust of vaults on the external side, tend to rotate. Generally, these are slow-moving phenomena, as demonstrated by the substantial absence of progression of the crack patterns detected by static monitoring, and reduced danger compared to instability of foundation structurer and/or soil. Based on all the above considerations, it is possible to follow a conservative approach by static monitoring, rather than providing with reinforcement of the vaulted structures. In fact, in the specific case, the principle of minimal intervention should be better achieved by the observation over the time of the crack propagation in order to address eventually the typology and extent of the restoration works.

## 3. Conclusions

The role of NDTs is crucial for the assessment and control of historical structures, provided that they are included in a structured diagnostic process, which moves from historical research and direct survey of the state of conservation and offers feedback to the qualitative, extended and indirect investigation through direct, targeted and localized tests and samples, with a view to limiting structural invasiveness and operational interference. In particular, the photogrammetric survey and the vibration detection for the verification of potential critical conditions for ancient artefacts are achieving, in recent times, wide application and full maturity, since they are based on expeditious procedures, with advantageous costs / benefits ratio. Specifically, the use of 2D image acquisition procedures for 3D model processing, in addition to the general advantages of using low-budget and user-friendly hardware and software tools for highly realistic documentation of the sites, results of particular interest in situations where it is necessary to have the geometric and material-constructive survey of places that are difficult to access, in confined spaces and / or in fragile contexts from a conservative point of view. This aspect, evident in the application of the case study for the underground spaces, may also emerge in other common situations in the architectural heritage, such as service rooms, rooms where it is not possible interrupt the operational functions, spaces in which there are disruptions that require to limit the stay of the operators as much as possible for safety reasons, spaces in which the anthropogenic presence can alter the microclimatic conditions of preservation of the artefacts. Similarly the vibration analysis allows the fast and effective acquirement of information about the dynamic response, for example in terms of accelerations, of structures, to be compared with suitable limit values affecting structural integrity and/or human sensitivity. In this way, crucial aspects can be evaluated, such as the discussed case of heavy traffic in proximity of ruined masonry walls. Besides, more in general, this type of approach provides a reliable and non-invasive experimental benchmark in order to calibrate theoretical and numerical models of historical constructions, so reducing the well-known inherent limits of the schematisations and simplifications of real objects.

## References

1. D.M McCann, M.C Forde, Review of NDT methods in the assessment of concrete and masonry structures, NDT & E International, Volume 34, Issue 2, 2001, Pages 71-84, ISSN 0963-8695.
2. Patricia Cotič, Zvonko Jagličić, Vlatko Bosiljkov, Validation of non-destructive characterization of the structure and seismic damage propagation of plaster and texture in multi-leaf stone masonry walls of cultural-artistic value, Journal of Cultural Heritage, Volume 15, Issue 5, 2014, Pages 490-498, ISSN 1296-2074.
3. Miroslav Sýkora, Dimitris Diamantidis, Milan Holický, Jana Marková, Árpád Rózsás, Assessment of compressive strength of historic masonry using non-destructive and destructive techniques, Construction and Building Materials, Volume 193, 2018, Pages 196-210, ISSN 0950-0618.
4. Rachel Martini, Jorge Carvalho, Nuno Barraca, António Arêde, Humberto Varum, Advances on the use of non-destructive techniques for mechanical characterization of stone masonry: GPR and sonic tests, Procedia Structural Integrity, Volume 5, 2017, Pages 1108-1115, ISSN 2452-3216.
5. Francisco J. Pallarés, Michele Betti, Gianni Bartoli, Luis Pallarés, Structural health monitoring (SHM) and Nondestructive testing (NDT) of slender masonry structures: A practical review, Construction and Building Materials, Volume 297, 2021, 123768, ISSN 0950-0618.
6. Naci Yastikli, Documentation of cultural heritage using digital photogrammetry and laser scanning, Journal of Cultural Heritage, Volume 8, Issue 4, 2007, Pages 423-427, ISSN 1296-2074,
7. Yahya Alshawabkeh, Mohammad El-Khalili, Eyad Almasri, Fadi Bala'awi, Amaal Al-Massarweh, Heritage documentation using laser scanner and photogrammetry. The case study of Qasr Al-Abidit, Jordan, Digital Applications in Archaeology and Cultural Heritage, Volume 16, 2020, e00133, ISSN 2212-0548.
8. María Higueras, Ana Isabel Calero, Francisco José Collado-Montero, Digital 3D modeling using photogrammetry and 3D printing applied to the restoration of a Hispano-Roman architectural ornament, Digital Applications in Archaeology and Cultural Heritage, Volume 20, 2021, e00179, ISSN 2212-0548.
9. Rosella Alessia Galantucci, Fabio Fatiguso, Advanced damage detection techniques in historical buildings using digital photogrammetry and 3D surface anlysis, Journal of Cultural Heritage, Volume 36, 2019, Pages 51-62, ISSN 1296-2074.
10. D Colapietro, A Fiore, A Netti, F Fatiguso, G C Marano, M De Fino, D Cascella, A Ancona, 'Dynamic identification and evaluation of the seismic safety of a masonry bell tower in the south of Italy', ECCOMAS THEMATIC CONFERENCE - COMPDYN 2013 - 4th International Conference on Computational Methods in Structural Dynamics and Earthquake Engineering, Proceedings – An IACM Special Interest Conference, pp 3459-3470, 2013.
11. A Fiore, G C Marano, 'Serviceability Performance Analysis of Concrete Box Girder Bridges Under Traffic-Induced Vibrations by Structural Health Monitoring: A Case Study', International Journal of Civil Engineering, Vol 16, No 5, pp 553-565, 2018.
12. A Cali, C Gentile, A Saisi, 'Vibration Testing and System Identification of a Monumental Building in Sabbioneta, Italy', Lecture Notes in Civil Engineering, Vol 128, pp 248 – 257, 2021.
13. G Lacanna, M Ripepe, M Coli, R Genco, E Marchetti, 'Full structural dynamic response from ambient vibration of Giotto's bell tower in Firenze (Italy), using modal analysis and seismic interferometry', NDT and E International, Vol 102, pp 9-15, March 2019.
14. A Doller, A Hondl, E Proksch, 'The construction of the Vienna Underground Railway and measures taken to protect St. Steven's Cathedral', Canada Institute for Scientific and Technical Information, National Research Council of Canada, NRC/CNR TT-1900, 1976.
15. ISO 2631-2:1989, 'Evaluation of human exposure to whole-body vibration. Part 2: Continuous vibration and shock-induced vibration in buildings (1 to 80 Hz)', International Organization for Standardization, 1989.

# Vibrational quality classification of metallic turbine blades under measurement uncertainty

Liangliang CHENG [1,2], Vahid YAGHOUBI [1,2], Wim V. PAEPEGEM [1], Mathias KERSEMANS [1]

[1] Mechanics of Materials and Structures (MMS), Ghent University, Technologiepark 46, B-9052,Zwijnaarde, Belgium
Email: liangliang.cheng@ugent.be; vahid.yaghoubi@ugent.be; wim.vanpaepegem@ugent.be; mathias.kersemans@ugent.be
[2] SIM program M3 DETECT-ION, Technologiepark 48 , B-9052 Zwijnaarde, Belgium

**Abstract**
Non-destructive testing on metallic turbine blades is a challenging task due to their complex geometry. Vibrational testing such as Process Compensated Resonance Testing (PCRT) has shown an efficient approach, which first measures the vibrational response of the turbine blades, and then employs a classifier to determine if the quality of the turbine blades are good or bad. Our previous work mainly concentrated on the development of Mahalanobis distance-based classifiers which are fed by the measured vibrational features (such as resonant frequencies). In practice, however, measurement errors could lead to a biased trained classifier, potentially resulting in the wrong quality classifications of the turbine blade. In this study, we investigate the classification problem of turbine blades under measurement uncertainty. For this, the concept of Interval Mahalanobis Space is employed, leading to the Integrated Interval Mahalanobis Classification system (IIMCS) classifier which has high robustness against measurement uncertainty.
The IIMCS employs Binary Particle Swarm Optimization (BPSO) to filter out those resonant frequencies which contribute most to the information in the system. A Monte Carlo Simulation scheme is employed to analyze the sensitivity of the resonant frequencies to the measurement uncertainty. This yields an indicator of reliability, indicating the confidence level of the final classification results. The developed IIMCS methodology is applied to an experimental case study of equiaxed nickel alloy first-stage turbine blades, showing good and robust classification performance.
**Keywords:** Mahalanobis distance, feature selection, classification, non-destructive testing, uncertainty propagation

## 1. Introduction

The Mahalanobis Taguchi System, termed MTS [1], is a pattern recognition method that incorporates the Mahalanobis Distance (MD) [2] designated to differentiate patterns between different groups, which has been applied not only to diagnose the system performance but also to optimize decisions in the system design and implementation.
The implementation of MTS involves two main components: Orthogonal Array (OA) employed to screen the significant variables and Signal-to-Noise Ratio (SNR) used to evaluate the measurement quality. Nevertheless, it serves as a standard diagnostic tool in many industrial fields, still being questioned by many scholars in the feature selection and threshold determination procedures [3-4]. To alleviate the shortcomings of the original MTS approach, the authors thereafter proposed the Mahalanobis Classification System (MCS) [5] and its integrated version (IMCS) [6], subsequently. More specifically, MCS employs Binary Particle Swarm Optimization (BPSO) [7] and PSO/Machine Learning tools (ML) [8-11] to filter out high dependable features and address the threshold determination issues, respectively. Despite the achieved promising outcomes using MCS, it still faces some challenges: a significant number of user-defined parameters and overfitting issues. To handle these challenges, an enhanced version of MCS was designed that incorporates threshold determination as part of the feature selection process. This Integrated Mahalanobis Classification System (IMCS) improved also the classification performance. For more details about MCS and IMCS, readers are referred to [5-6].

However, the presence of measurement uncertainty as a natural attribute in conducting experiments results in biased classifiers in the source data that were not considered in either MCS or IMCS. The present work presents a simple, yet effective, manner to account for measurement uncertainty by using the concept of Interval Mahalanobis Distance (IMD) [12]. This IMD is integrated into the classification scheme, resulting in the Integrated Interval Mahalanobis Classification System (IIMCS). Apart from improving the classification procedure, it also provides an effective indicator for classification reliability.

## 2. Integrated interval Mahalanobis classification system

Despite the demonstrated favorable performance of (I)MCS for classifying metal parts based on their vibration responses, it does not take into account possible measurement uncertainties and provide an assessment of the reliability of the classification results. To achieve this, the (I)MCS classification framework was extended by merging different concepts, resulting in the Integrated Interval Mahalanobis Classification System IIMCS. The general scheme of the IIMCS algorithm is shown in Figure 1.
The main novelties in IIMCS are the implementation of Monte Carlo Method [13] ( termed MCM in Figure 1) to infer the Interval MD and an indicator to point out the classification reliability.
The IMD has been defined as equation (1)

$$\boldsymbol{IMD}_k = [MD_k^1, MD_k^2, \dots, MD_k^m] \tag{1}$$

In which $m$ and $k$ stand for the MCM total runs and sample index.
In practical applications, the impacts of data uncertainty and learner stability can be dispersed on the final classification results. Benefiting from the concept of IMD, classification reliability can be inferred and considered as an effective tool to indicate the reliability of classification results. The classification reliability is indicated as $r$, ranges from [Good, Bad] = [-1,1]. The plus-minus sign of reliability values indicates the sample classified as Bad (-) and Good (+). And the absolute values of the reliability imply the possibility to classify the sample as Bad or Good. The greater the absolute value of the reliability, the more confident the prediction.
Thus, the reliability of classification is defined as

$$r = \begin{cases} -p^{Good} & if\ p^{Good} > p^{Bad} \quad r \in [-1, -0.5) \\ p^{Bad} & if\ p^{Good} < p^{Bad} \quad r \in (0.5, 1] \\ 0 & if\ p^{Good} = p^{Bad} \quad r = -0.5, 0.5 \end{cases} \tag{2}$$

In which $p^{Good}$ and $p^{Bad}$ are used to represent the assigned probability to the sample being classified as Good and Bad, respectively. The following example in Figure.1 illustrates the concept of classification reliability.
It is assumed that the logarithmic of chi-squared distributed variables is approximately Gaussian distributed and the scaled IMD approximately then corresponds to a Gaussian distribution [5]. Figure 1 indicates that the probability of being classified as Bad and Good is 90% and 10%, respectively, according to equation (2), drawing a conclusion that this part is sorted as Bad with a classification reliability of 0.9 ($r = p^{Bad} = 0.9$).

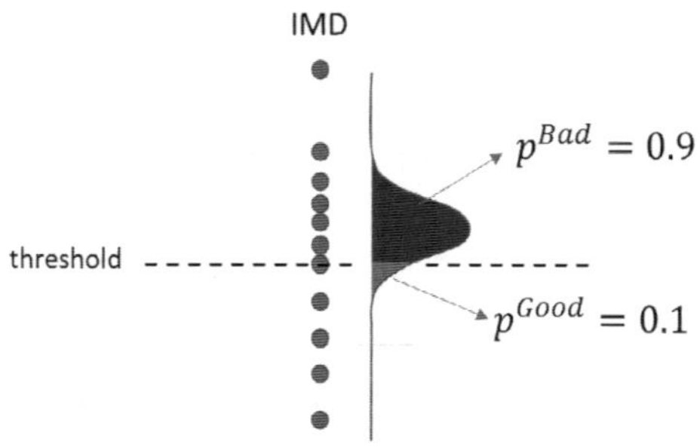


Figure.1 The illustration of classification reliability

Hence, this approach provides an opportunity to investigate how reliable the classification results are.

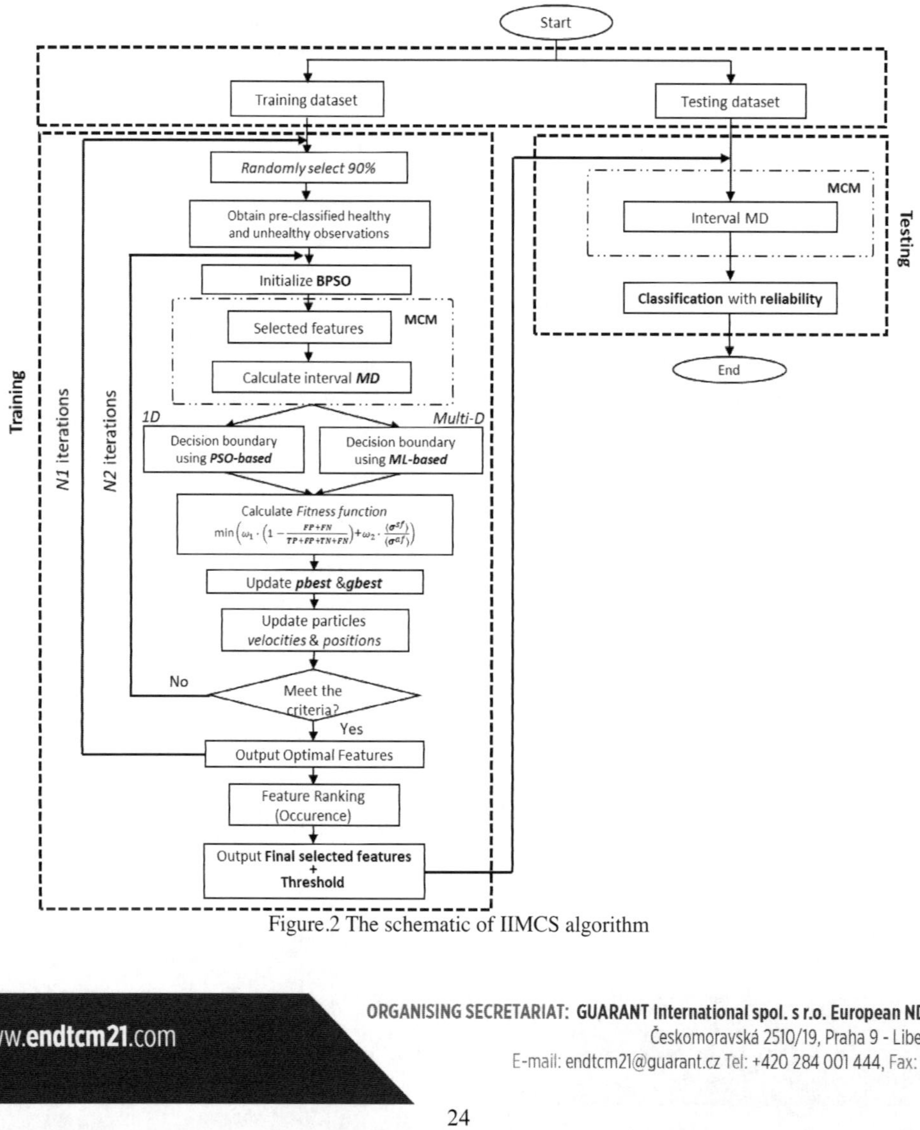


Figure.2 The schematic of IIMCS algorithm

In short, the flowchart of IIMCS in Figure.2 can be briefly described as: The original dataset consists of *n* samples and *p* features, which are first randomly divided into training and test datasets. Using the BPSO optimizer with *N2* iterations, the optimal subset of features is searched in the *m*-run MCM framework to maximize the classification accuracy of the training dataset. Furthermore, the whole operation will be run *N1* times to further filter out the most reliable features based on their frequency of occurrence. Note that the threshold determination in this study centers only on the one-dimensional MD space, and therefore the PSO optimizer is utilized to exploit the optimal threshold on the basis of the resulting IMD values rather than the single point MD values in both MCS and IMCS.

## 3. A case study on metallic turbine blades

This section shows a case study of an equiaxed nickel alloy primary turbine blade with a generic model and its sectional CAD drawing displayed in Figure 3. A total of 232 blades consisting of 193 healthy and 39 defective blades are available. These defective blades contain several types of defects, such as microstructural changes due to high temperatures; airfoil cracking; intergranular erosion (corrosion); thin walls due to casting; maintenance and repair operations, and service wear.

Figure.3a Generic model of the turbine blade

Figure.3b Sectional drawing of the turbine blade

Vibrant Corporation conducted vibration testing of turbine blades in the frequency range of [3, 38] kHz utilizing compensated resonance testing (PCRT) techniques [14]. From the vibration response, 15 resonance frequencies $f_1 \sim f_{15}$ were identified for further analysis. To assess the potential measurement uncertainty or repeatability of the measurements, 31 repeated measurements were performed on the nominal turbine blades.

Figure.4a and Figure.4b present the IMD results based on a complete set of 15 features $f_1 \sim f_{15}$ and the optimized subset of features $(f_2, f_4, f_{14}, f_{15})$, respectively. For better visualization, the Scaled MD [5], which is the log-normalized MD, is used. It is evident from Figure.4 that the margin of IMD values is significantly shrunk via the adoption of the optimal feature subset $(f_2, f_4, f_{14}, f_{15})$.

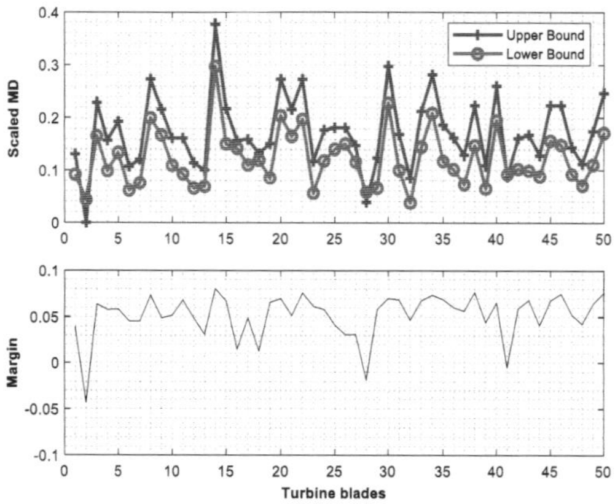

Figure.4a IMD based on the whole fifteen features

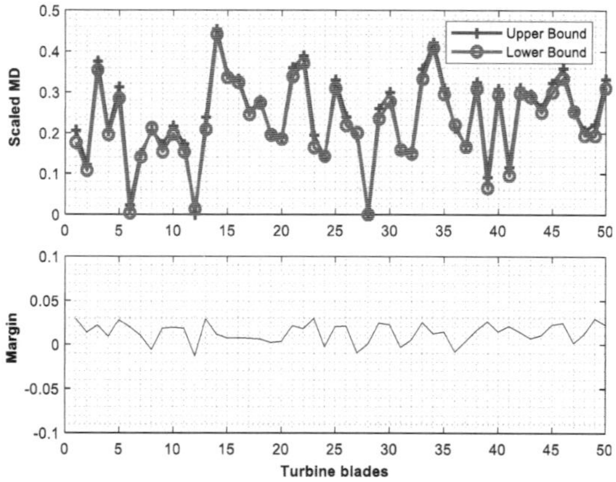

Figure.4b IMD based on the selected four features

In addition to having a narrower range of IMD values, the reliability of each test blade shown in Figure 5 also facilitates the determination of the health status of the test turbine blades. Two misclassifications on the testing dataset are observed, resulting in an accuracy of 97.85%. Of interest, the turbine blade 87 is falsely categorized as healthy, but only with a reliability value of $r$= -0.78.

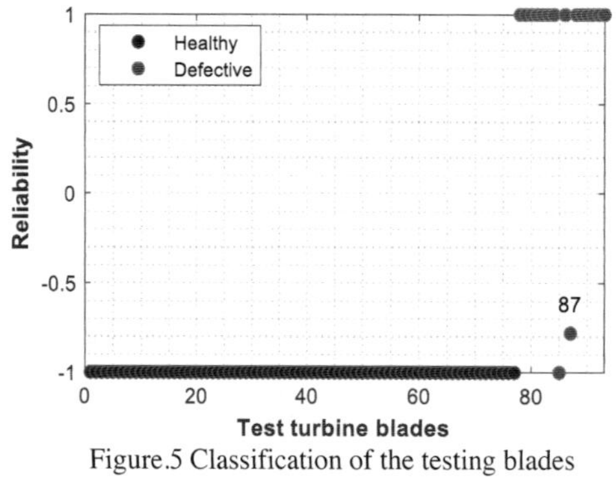

Figure.5 Classification of the testing blades

## 4. Conclusions

An Integrated Interval Mahalanobis Classification System IIMCS is proposed for classifying complex metal parts based on their broadband vibration response. The IIMCS classifier is proposed to address the challenges in dealing with measurement uncertainties in the context of the Mahalanobis-Taguchi System MTS. IIMCS employs a Monte Carlo method framework to propagate the measurement uncertainty into the Mahalanobis distance space MD to obtain the interval MD space and assigns an indicator to each test part intending to evaluate the classification reliability. The proposed IIMCS is demonstrated in an experimental case study of an equiaxed nickel alloy first stage turbine blade, showing satisfactory performance.

**Acknowledgements**

The authors acknowledge the ICON project DETECT-ION (HBC.2017.0603) which fits in the SIM research program MacroModelMat (M3) coordinated by Siemens (Siemens Digital Industries Software, Belgium) and funded by SIM (Strategic Initiative Materials in Flanders) and VLAIO (Flemish government agency Flanders Innovation & Entrepreneurship).

Vibrant Corporation is gratefully acknowledged for providing anonymous datasets of the turbine blades.

**References**

1. Taguchi G, Jugulum R. 'The Mahalanobis-Taguchi strategy: A pattern technology system'. John Wiley & Sons; 2002 May 23.
2. McLachlan GJ. 'Mahalanobis distance'. Resonance. 1999 Jun 1;4(6):20-6.
3. Woodall WH, Koudelik R, Tsui KL, Kim SB, Stoumbos ZG, Carvounis CP. 'A review and analysis of the Mahalanobis—Taguchi system'. Technometrics. 2003 Feb 1;45(1):1-5.
4. Pal A, Maiti J. 'Development of a hybrid methodology for dimensionality reduction in Mahalanobis–Taguchi system using Mahalanobis distance and binary particle swarm optimization'. Expert Systems with Applications. 2010 Mar 1;37(2):1286-93.
5. Cheng L, Yaghoubi V, Van Paepegem W, Kersemans M. 'Mahalanobis classification system (MCS) integrated with binary particle swarm optimization for robust quality classification of complex metallic turbine blades'. Mechanical Systems and Signal Processing. 2021 Jan 1;146:107060.
6. Cheng L, Yaghoubi V, Van Paepegem W, Kersemans M. 'Quality inspection of complex-shaped metal parts by vibrations and an integrated Mahalanobis classification system'. Structural Health Monitoring. 2018:1475921720979707.
7. Kennedy J, Eberhart RC. 'A discrete binary version of the particle swarm algorithm'. In1997 IEEE International conference on systems, man, and cybernetics. Computational cybernetics and simulation 1997 Oct 12 (Vol. 5, pp. 4104-4108). IEEE.
8. Kennedy J, Eberhart R. 'Particle swarm optimization'. InProceedings of ICNN'95-international conference on neural networks 1995 Nov 27 (Vol. 4, pp. 1942-1948). IEEE.
9. Yaghoubi V, Cheng L, Van Paepegem W, Kersemans M. 'A novel multi-classifier information fusion based on Dempster–Shafer theory: application to vibration-based fault detection'. Structural Health Monitoring. 2020 Dec 4:14759217211007130.
10. Yaghoubi V, Cheng L, Van Paepegem W, Kersemans M. An ensemble classifier for vibration-based quality monitoring. Mechanical Systems and Signal Processing. 2022 Feb 15;165:108341.
11. Cheng L, Yaghoubi V, Paepegem WV, Kersemans M. 'On the Influence of Reference Mahalanobis Distance Space for Quality Classification of Complex Metal Parts Using Vibrations'. Applied Sciences. 2020 Jan;10(23):8620.
12. Chang Z, Cheng L, Cui L. 'Interval choquet fuzzy integral multi-attribute decision-making method based on mahalanobis taguchi system'. Control Decis. 2016;31(1):180-6.
13. Rubinstein, R. Y., & Kroese, D. P. 'Simulation and the Monte Carlo method' (Vol. 10). John Wiley & Sons. 2016.

14. Heffernan JV, Jauriqui L, Biedermann E, Mayes A, Livings R, Goodlet B, Mazdiyasni S. 'Process compensated resonance testing models for quantification of creep damage in single crystal nickel-based superalloys'. Materials Evaluation. 2017 Jul 1;75(7):941-52.

# Acoustic emission based damage monitoring to study fracture in cementitious mortars under fatigue loading

Nuhamin E. DERESSE [1], Charlotte VAN STEEN [1], Mina SAREM [1], Stijn FRANÇOIS [1], Els VER-STRYNGE [1],

[1]Department of Civil Engineering, KU Leuven
Leuven, Belgium; nuhamineshetu.deresse@kuleuven.be, charlotte.vansteen@kuleuven.be, mina.sarem@kuleuven.be, Stijn.francois@kuleuven.be, els.verstrynge@kuleuven.be

**Abstract**
In this work, acoustic emission (AE) sensors are used to detect micro-cracking in mortar cylinders subjected to monotonic and step-wise fatigue loads, in a Brazilian splitting test setup. The difference in damage progression between monotonic and fatigue tests was analyzed by using AE monitoring. The rate of AE events in monotonic tests was initially low, but there is a sudden rise of AE events with high amplitude and energy prior to failure. On the other hand for fatigue tests, three damage stages were identified. In addition, the width of a fracture process zone (FPZ) measured with AE, was found to be higher for the fatigue tests. Finally, crack types throughout the fracture process are classified using AE parameter analysis. In both tests, initially, AE due to tensile cracks dominated. However, after the major crack has formed, more shear noises were observed.

**Keywords:** Acoustic emission, mortar, Brazilian splitting test, fatigue, fracture process zone, crack types

## 1. Introduction

Fatigue is the deterioration and fracture of materials due to cyclic loads (Suresh, 1991). When a load that is not large enough to cause failure in static application causes failure with a repetitive application, it is regarded as fatigue failure (Tait, 1984). When low amplitude loads are applied on a brittle material like concrete, initially the material seems intact at the macroscopic scale. However microscopically, due to inherent flaws, there are areas of stress concentration resulting in micro-crack propagation. Even if this is a minor propagation, it will effectively change the mechanical properties of the material. Thus, due to the repetitive occurrence of these micro-crack propagations, they can join to form a significant change to the bulk mechanical properties. Fatigue and the factors that affect fatigue in concrete have been the subject of research since the early 20th century. However, with the advancement of non-destructive testing technologies and knowledge and application of fracture mechanics in concrete, more understanding of fatigue phenomena can now be obtained.

Fatigue damage in concrete is associated with micro-damage, which requires a sensitive detection techniques like the Acoustic Emission (AE), which is instrumental for monitoring damage progression. In early work, Tait (1984) used the AE technique to monitor damage progression in mortars subjected to compressive fatigue loading. The research found that damage in fatigue develops in three stages; the initial increase in AE activity is followed by a linear rate of growth, and finally there is a steep rise of AE events prior to failure. More recently, Keerthana and Kishen (2020) used AE monitoring to compare the formation of the fracture process zone (FPZ) between monotonic and fatigue three-point bending tests of plain concrete beams. The research concluded that in monotonic loading, a more localized inelastic zone in front of the fracture front can be found. However, under fatigue, more isolated and dispersed micro-cracks were observed.

In this paper, an experimental program is designed to study and relate the fracture progression during fatigue and monotonic loading in Brazilian splitting tests by means of AE monitoring. The aim is to compare observed micro-fracture progress between both loading schemes and investigate whether literature findings on compression and three-point bending tests can be confirmed for a Brazilian splitting test setup by means of AE analysis. For the working principle of the AE technique, the reader is referred to the available literature (Grosse, Ohtsu, Aggelis, & Shiotani, 2022).

## 2. Experimental procedure

Mortar cylinders were made in accordance with EN 196-1. The mortar composition and sample geometry are shown in **Table 1** and **Table 2** respectively. The composition is tested according to EN 196-1, resulting in 20.65 MPa and 0.71 MPa mean and standard deviation of compressive strength respectively and 3.56 MPa and 0.33 MPa mean and standard deviation of flexural strength respectively.

**Table 1 Mortar composition**

| Material | (kg/$m^3$) |
|---|---|
| CEM II/B (S-L) 32.5 N | 533 |
| Sand 0/2 | 1599 |
| Water | 266.5 |
| W/C | 0.5 |

**Table 2 Specimen geometry**

| | | No of samples | |
|---|---|---|---|
| Diameter (mm) | Length (mm) | Monotonic | Fatigue |
| 106 | 106 | 5 | 2 |

Monotonic and step-wise fatigue Brazilian splitting tests are performed on a universal testing machine (Shimadzu AG-XMST-X/X-TYPE, having a maximum capacity of 100 kN). The experimental test setup along with the AE sensor layout is shown in Figure 1. During the monotonic and step-wise fatigue tests, six piezoelectric broadband sensors (Digital Wave B1025) with a 50 to 2000 kHz frequency range are mounted on the sample surface with a thin layer of hot-melt glue to monitor the AE activities. The sensors were connected to preamplifiers with 34 dB gain, and subsequently to a Vallen AMSY-6 acquisition system with six AE channels. The acquisition system frequency filter was set between 25 kHz to 850 kHz. The threshold was set at 40 dB, a value determined by monitoring the environmental noise of the laboratory.

During a Brazilian test, tensile cracks propagate in a rather abrupt way, which requires a high rate of data acquisition for the emitted acoustic emissions. Therefore, the maximum possible sampling rate for Vallen AMSY-6 (ASIP-2/S), i.e. 10 MHz, is used. Further, a Rearm time (RAT) and duration discrimination time (DDT) of 170 μs are adopted for the tests.

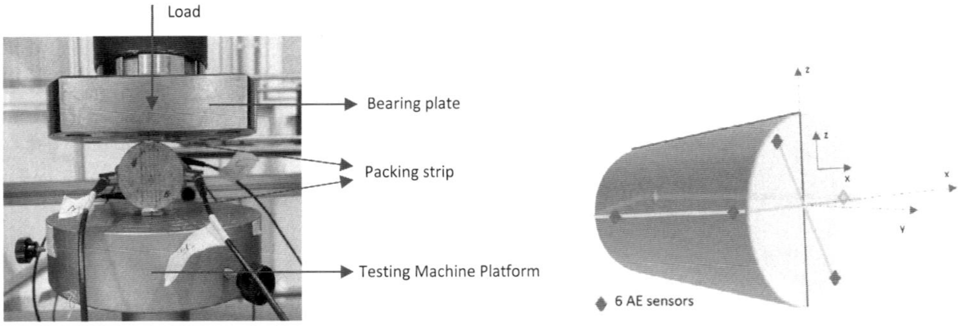


Figure 1 Experimental setup for Brazilian splitting test (left) and AE sensors configuration (right)

## 3. Results and discussion

In accordance with EN 12390-6, the monotonic tests were force-controlled, with a loading rate of 666 N/s. The fatigue loading scheme is shown in Figure 3, which is a displacement-controlled loading rate with upper and lower load limits. For each of the steps, depending on the sample geometry, an adoptive displacement rate that would generate a similar loading frequency was incorporated.

### *3.1. AE parameter analysis*

Firstly, the cumulative AE event count with the related event amplitude and energy is discussed for a representative sample. For the monotonic loading, even though there is AE activity from the beginning, there is a significant increase in AE event rate near the peak load, and high amplitude and high energy events do not occur before approaching the failure load, as seen in Figure 2. On the other hand, for fatigue tests, three damage stages were identified, see Figure 3 (left). There is an initial increase of AE activity during first load application which is related to concentrated compressive stress near the contact surface (stage 1). Following this, there is a region of lower increase in AE activity (stage 2). At the final stages, cracks initiate and propagate with each load cycle, which causes a steep increase in AE event rate right before failure (stage 3). In stage 3, AE events have a significantly higher amplitude and energy. This is shown in Figure 3 (middle) and (right). The slower increase in AE activity during fatigue cycles in contrast to monotonic loading indicates progressive and cumulative damage.

Also similar to Keerthana and Kishen (2020), more low-energy AE events are found in fatigue loading than in monotonic loading.

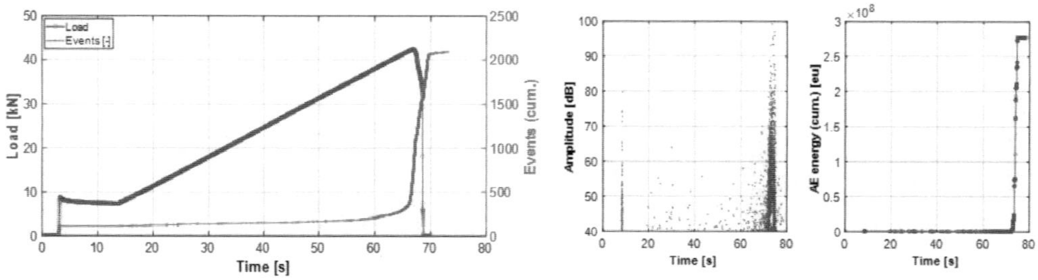


Figure 2 Load versus time and cumulative AE event versus time (left), amplitude versus time (middle) and cumulative AE energy versus time (right) of a monotonic test

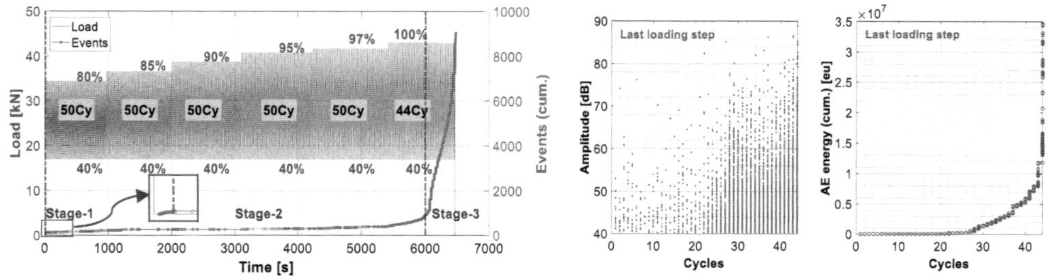


Figure 3 Load versus time and cumulative AE event versus time curve (left), amplitude versus time (middle), and cumulative AE energy versus time (right) of a step-wise fatigue test

### *3.2. AE-based fracture process zone (FPZ) analysis*

Otsuka and Date (2000) defined the FPZ as the AE event region where 95% of the total AE energy is included. Hadjab, Thimus, and Chabaat (2007) and Alam, Saliba, and Loukili (2014) used a plot of the distribution of located AE events along the horizontal direction of the sample. The width of FPZ is taken as the width of the plot at 20% of the maximum located cumulated AE events. This latter approach is also adopted here.

The AE-based FPZ are shown in Figure 4 for a sample under fatigue (left) and monotonic (right) loading. It can be observed that the estimated width of FPZ for the step-wise fatigue sample (28.9 mm) is larger than for the monotonic test (18.1 mm). When comparing monotonic and fatigue three-point bending tests of plain concrete beams, Keerthana and Kishen (2020) reported that even though there is a well-defined fracture zone for monotonic loading, micro-cracks were found to be isolated and scattered for fatigue loading, which led to the conclusion that the FPZ in their definition does not exist for fatigue loading. Here, both loading schemes generate localized cracks mainly thanks to the test setup (i.e. Brazilian splitting). Yet similar to Keerthana and Kishen (2020) more scattered micro-events are still observed for step-wise fatigue loading, resulting in the larger width of the FPZ. This can be due to the highly brittle nature of the fracture in the monotonic test, which did not allow enough AE events to be captured and located. However, it is also likely related to the difference in the fracture process between the two tests.

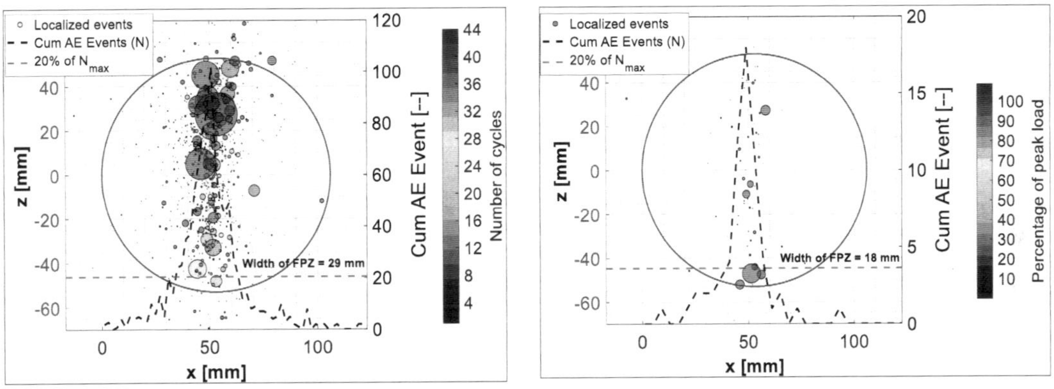


Figure 4 AE-based width of FPZ estimation for step-wise fatigue (left) and monotonic (right) loading schemes

### *3.3. Average Frequency (AF) and Rise angle (RA)*

Finally, Rilem TC 212-ACD recommendation (Ohtsu, Isoda, & Tomoda, 2007) is used to classify the type of active cracks. The average frequency (AF) and rise angle (RA) are calculated and used to identify AE signals due to tensile cracks from shear cracks and mixed source types. AF and RA are calculated using equations 1 and 2.

$$\text{Raise angle (RA)} = \frac{\text{Rise time}}{\text{Maximum amplitude}} \left[\frac{mV}{s}\right] \quad (1)$$

$$\text{Average frequency (AF)} = \frac{\text{AE ring-down count}}{\text{Duration}} \; [kHz] \quad (2)$$

Firstly, a sample is cracked monotonically using the Brazilian splitting test which causes failure mainly by tensile cracking. To verify this through AE source classification, first a sample is cracked by the splitting test. Then the cracked sample is taken and friction noise is generated manually by rubbing both halves together. The AF and RA results of these two tests are shown in Figure 5. As it can be seen, there is a clear distinction between the AE signals generated by the tensile crack and friction-induced noise. The results are in line with the Rilem recommendation, i.e. AE signals generated from tensile cracks have shorter duration, higher counts, and shorter rise time, resulting in higher AF and lower RA.

The localized events shown in Figure 5 formed two distinctive clusters. The RA to AF proportion set at 0.56 was found to isolate the clusters. However, when taking 0.1 as the slope, which is the value suggested by Ohtsu et al. (2007), the clusters can no longer be identified individually. The value 0.56 is estimated by trial and error in an attempt to isolate the two clusters in both plots and is shown on Figure 5. These proportional ratio on RA and AF plot that can divide AE coming from tensile and shear crack is still not fixed to specific criteria and needs to be estimated for each material and structural condition (Ohtsu et al., 2007). From the observation within the studied data sets, it can be concluded that the AE events found above 0.56 are only due to tensile cracks, those found below are mainly due to shear/friction noise.

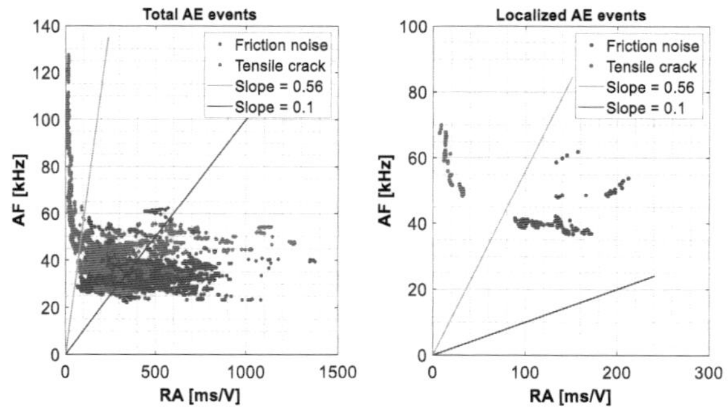


Figure 5 AF and RA plots of tensile crack and friction noise with the total events (left) and localized events (right).

RA and AF values found from monotonic and step-wise fatigue tests are shown in Figure 6 and Figure 7 respectively. Generally, for both cases, most of the AE events are found under tensile crack classification. For the monotonic loading, see Figure 6, some AE events at the beginning are classified as shear crack. This is related to the concentrated compression stress at the contact surface. Nevertheless, most of the AE events occurring during the pre-peak (70-95%) and peak (95-100%) loading period are classified as tensile cracks. However, the AE events during the post-peak period (100-0%) are due to a mix of tensile cracks and shear noise. This can be due to the friction noise that occurs along the developed major cracks after the sample has failed. For selected cycles of the fatigue test, the AF and RA plots are shown in Figure 7. At the earlier cycles, the majority of the events are classified as tensile cracks. However, as cracks are formed, there are more and more events falling into the shear noise category, and eventually in the very last cycle more emissions due to shear are noticed. This again can be due to the friction noise caused when cracked parts slide on one another.

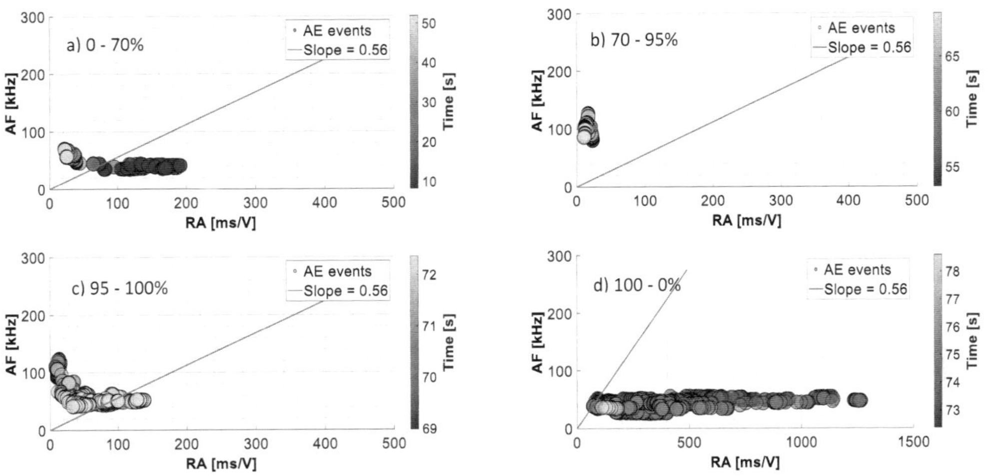


Figure 6 AF and RA plots of AE generated during a monotonic test; a) to c) are plots in pre-peak load period divided with percentage of peak load; d) is a plot in the post-peak period.

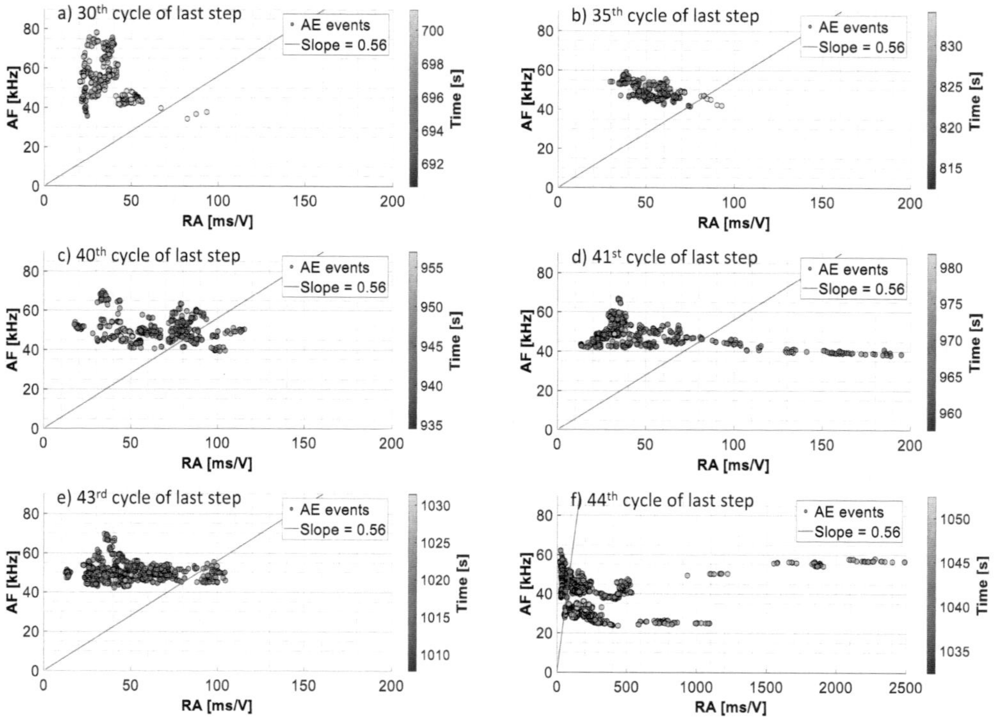


Figure 7 AF and RA plots of step-wise fatigue test (each plot is related to a specific cycle within the last loading step).

## 4. Conclusions

This paper discusses the application of AE monitoring to analyse the fracture process under monotonic and step-wise fatigue loading on brittle mortars in a Brazilian splitting test setup.

Initially, for monotonic loading, there are a small number of low amplitude and energy AE events. Conversely, AE events with high energy and amplitude start to increase at a higher rate just prior to failure. On the other hand for fatigue tests, three damage stages were observed. Stage-1 corresponded to the initial increase of AE activity related to the compression near the contact surface due to stress concentration. In stage-2, there is a region of slower increase in AE activity, and finally at stage-3, cracks initiate and propagate with each load cycle, which causes an increase of AE events with a high amplitude and energy. The more progressively increasing AE events in fatigue tests as opposed to the sudden increase in monotonic test indicates, damage caused by fatigue tests is more progressive and develops cumulatively.

The width of FPZ in monotonic and in fatigue tests is estimated from the located AE events. It was found that due to a more scattered micro-events, the width of the FPZ during fatigue is larger than that of a monotonic loading test.

Finally, classification of active cracks throughout the fracture process is made using RA and AF analysis of the AE events. It was noticed that for most of the test duration, the emissions are due to tensile cracks. Yet, close to the failure stage and in the post-peak region when crack surfaces have fully developed, there are more shear/friction noises.

## Reference

Alam, S. Y., Saliba, J., & Loukili, A. (2014). Fracture examination in concrete through combined digital image correlation and acoustic emission techniques. *Construction and Building Materials, 69*, 232-242. doi:https://doi.org/10.1016/j.conbuildmat.2014.07.044

Grosse, C. U., Ohtsu, M., Aggelis, D. G., & Shiotani, T. (2022). *Acoustic Emission Testing* (Vol. Second Edition): Springer.

Hadjab, H. S., Thimus, J. F., & Chabaat, M. (2007). The use of acoustic emission to investigate fracture process zone in notched concrete beams. *Current Science, 93*(5), 648-653.

Keerthana, K., & Kishen, J. M. C. (2020). Micromechanics of fracture and failure in concrete under monotonic and fatigue loadings. *Mechanics of Materials, 148*, 103490. doi:https://doi.org/10.1016/j.mechmat.2020.103490

NBN EN 196-1:2016: Methods of testing cement. Determination of strength. (2016). In: NBN.

NBN EN 12390-6:2009: Testing hardened concrete. Tensile splitting strength of test specimens. (2010). In: NBN.

Ohtsu, M., Isoda, T., & Tomoda, Y. (2007). Acoustic emission techniques standardized for concrete structures. *Journal of Acoustic Emission, 25*(2007), 21-32.

Otsuka, K., & Date, H. (2000). Fracture process zone in concrete tension specimen. *Engineering Fracture Mechanics, 65*(2), 111-131. doi:https://doi.org/10.1016/S0013-7944(99)00111-3

Suresh, S. (1991). *Fatigue of materials*: Cambridge : Cambridge University press.

Tait, R. B. (1984). *Fatigue and fracture of cement mortar.* University of Cape Town, Retrieved from http://hdl.handle.net/11427/23365 Available from University of Cape Town OpenUCT database.

Verstrynge, E. (2019). 10 - Acoustic emission testing. In B. Ghiassi & P. B. Lourenco (Eds.), *Long-term Performance and Durability of Masonry Structures* (pp. 287-307): Elsevier.

# Development of an Ultrasonic Testing System for Press Connectors of Railway Catenary Wires

Thomas HERZOG [1], Frank SCHUBERT [1], Stephan HEILMANN [1], Jürgen MICHAUK [1], Susan WALTER [1], Henning HEUER [1]

[1] Systems for Testing & Analysis, Fraunhofer Institute for Ceramic Technologies and Systems IKTS, Dresden, Germany; Phone: +49 351 88815-523, Fax +49 351 88815-509; e-mail: frank.schubert@ikts.fraunhofer.de, thomas.herzog@ikts.fraunhofer.de, stephan.heilmann@ikts.fraunhofer.de, juergen.michauk@ikts.fraunhofer.de, susan.walter@ikts.fraunhofer.de, henning.heuer@ikts.fraunhofer.de

**Abstract**
This paper describes the development of an ultrasonic testing system for press connectors of railway catenary wires. With this system it is possible for the first time to detect the formation of cracks at an early stage. By means of a test gripper which is adapted to the hexagonal cross section of the press connector, several ultrasonic probes are pressed onto the crimped surfaces. With an appropriate wedge the ultrasound waves are guided to the inner cracks in order to generate echo signals in reflection and to ensure a stable coupling control. During inspection of new and existing installations, potentially damaged press connectors can be quickly detected and replaced if necessary. The system is based on IKTS' proprietary PCUS® *pro* electronics and has been successfully evaluated in first field tests. The prototype consists of six ultrasonic probes integrated into the gripping pliers so that a complete compression joint can be tested without rotating the gripper.

**Keywords:** Ultrasonic testing system, railway, press connector, catenary wire, compression joint, crack detection

## 1. Introduction

In the field of electric railway technology, tension-proof press connectors made of copper-based alloys are used to connect the support cables mechanically and electrically (Fig. 1, on the left). They have to carry static tensile forces up to 10.8 kN permanently, withstand the dynamic vibration load that occurs when a train passes through and guarantee a service life of 30 years under the harsh, climatic conditions at the place of use. In addition, the electric current needs to be guided safely and reliably without an excessive heating of the connection. Consequently, after pressing the copper sleeve to a hexagonal cross section to mechanically fix the cable ends inside, the electrical connection resistance must be as low as possible. A major problem during the pressing of the copper sleeves is the initiation of cracks in the corners of the hexagonal cross section (Fig. 1, on the right). These cracks start at the inner part of the sleeve and may grow to the outside during operation.

During regular inspections, the press connections can be tested only from the outside by visual inspection so far [1]. Therefore, only when the crack has already grown to the outside, it can be determined as a potential connector failure. The aim of the present work was to develop an ultrasonic testing system capable of detecting the formation and growth of cracks at an earlier stage. This would allow to identify defects not only during operation but also immediately after the initial pressing of the sleeves in the frame of repairs or reinstallations of suspension cable sections.

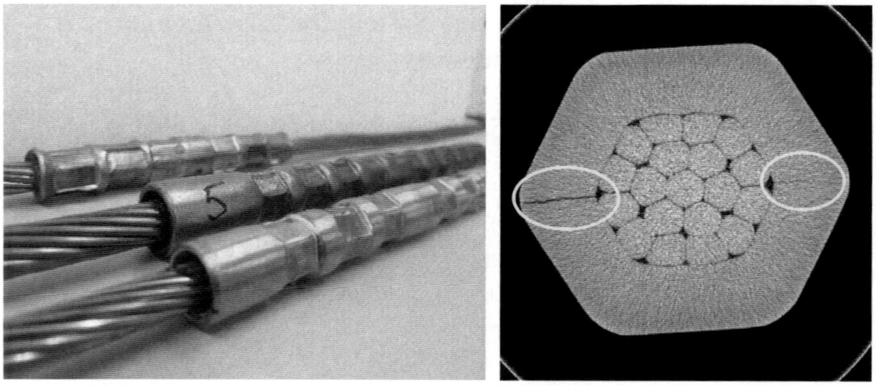

Figure 1. Copper-based press connectors (on the left) and X-ray CT of typical corner cracks in the hexagonal cross section caused by the pressing of the sleeves during cable fixation (on the right).

## 2. Development of a phased array prototype

### *2.1 Concept*

In a first stage a phased array system for inspection of the press connectors was developed. Its main principle is shown in Fig. 2 (on the left). A single 10 MHz linear phased array probe is integrated in a gripper and pressed onto the crimped surfaces of the press connector by means of a 9 mm high zero-degree Rexolite® wedge.

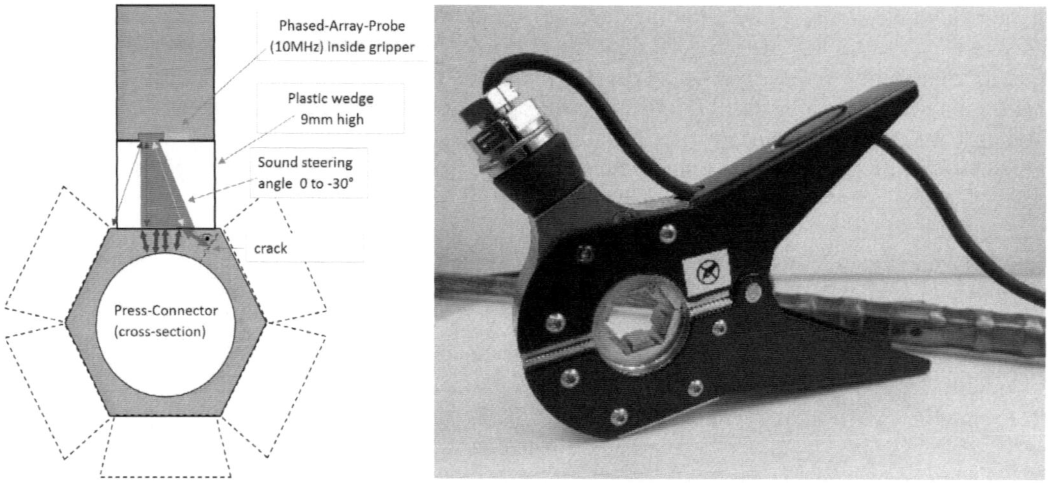


Figure 2. Concept and implementation of phased array inspection for press connectors.

The ultrasonic waves can be generated with incidence angles between 0° and +30° in order to capture echo signals of inner cracks from the right edge of the accessible surface (small green arrow in the left picture of Fig. 2) as well as the echo from the inner cylindrical borehole of the press connector (red arrows). The latter is used to ensure a reliable coupling control between wedge and press connector. The maximum swivel angle of 30° is a direct result of the sixfold cross-sectional geometry of the press connector and the compressional wave speeds of wedge ($c_P$ = 2335 m/s) and connector ($c_P$ = 4697 m/s), respectively. The main contribution from a crack in the corner is expected at 26° incidence angle. This corresponds to an emergent angle of 60° in the copper alloy.

### *2.2 Implementation*

For the prototype, a 10 MHz phased array probe PA10L16 with 16 elements (Olympus) was integrated in a ruggedly designed mechanical gripper (Fig. 2, on the right). It can be opened by the human operator by pressing the two handle pieces together. The closing is automatically done by an integrated spring mechanism when the compressive force on the handles is untightened. The acoustic coupling was realized by an automatic water supply system. The ultrasound signals are controlled by a PCUS® *pro* Array electronics (Fig. 3, on the right), i.e. the phased array version of the PCUS® *pro* ultrasound line developed at IKTS (see [2] for the latest development). It is integrated in a switch cabinet with industrial PC, monitor, hand-held control, and GPS system, everything placed on the same inspection train (Fig. 3). A customized data acquisition, processing, and evaluation software, the PCUS® *pro* Lab, was also developed during the course of the project. It ensures a reliable and stable inspection according to national and international norms and standards. A typical screenshot of the graphical user interface during inspection is depicted in Fig. 4.

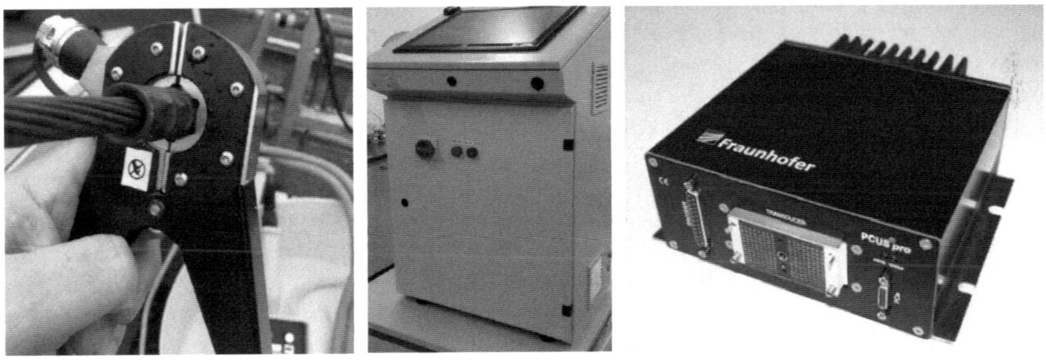


Figure 3. Field test with the phased array testing equipment located on an inspection train.

### *2.3 Field tests and first results*

A successful field test of the whole equipment was performed on a railway line (see Fig. 3). During the measurements it was demonstrated that real cracks in the press connectors can be detected reliably (Fig. 4). They were found predominantly in the direction of 26° incidence angle (or 60°

emergent angle) as expected due to the specific geometry of the cross-sectional area of the press connectors shown in Fig. 1. Moreover, a stable coupling control was achieved by monitoring the 0° echo from the inner connector borehole.

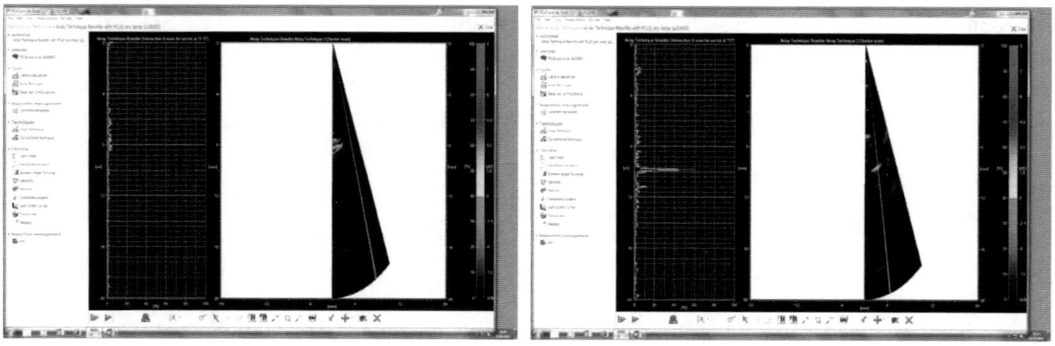

Figure 4. Typical result of a phased array sector scan. The principal direction of expected crack echoes is at 26°. On the left: measurement without defect (only the borehole echo at 0° is visible); on the right: measurement with defect echo along the 26° direction (white line).

### *2.4 Drawbacks of the phased array approach*

Despite its convincing performance in detecting cracks in the press connectors, the phased array approach reveals two serious disadvantages. One drawback is the fact that a complex and expensive phased array electronics together with an appropriate phased array transducer is needed although only two different angles of incidence (0° and 26°) are relevant for inspection. Moreover, with the current system only one of the six different contact areas of the press connector can be inspected at the same time. In order to inspect all six contact areas, the transducer needs to be relocated five times by opening, rotating, and closing the mechanical pliers each time. This leads to a high expenditure of time and an increased risk for missing relevant crack echoes. The integration of six phased array transducers in the same gripper was no option due the limited space and the significantly higher costs. Therefore, together with the customer it was decided to change the gripper to a set-up with six conventional ultrasonic probes being integrated into the pliers. This would allow the complete inspection of a press connector at the same time without rotating the gripper. Moreover, it was decided to replace the water coupling by application of a higher viscous coupling agent. The development of this second and final prototype is explained in the next chapter.

## 3. Development of a single-channel system with six conventional probes

### *3.1 Gripper concept*

Instead of using a multi-channel system with a single phased array transducer, a new design based on six conventional normal probes and a single-channel electronics with multiplexer was designed. The principal transducer set-up and a CAD drawing of the new inspection gripper are shown in Fig. 5.

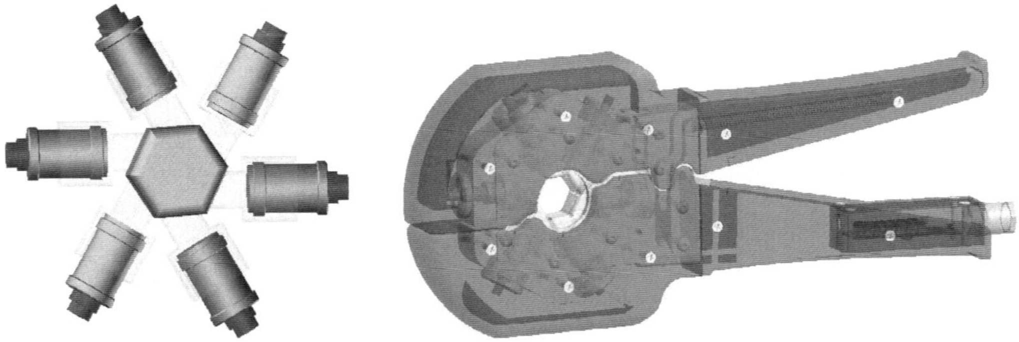

Figure 5. Transducer configuration with six conventional normal probes and 26° wedges (on the left) and a CAD drawing of the corresponding inspection gripper (on the right). The multiplexer is integrated in the lower grip of the pliers.

### *3.2 Coupling control*

While the phased array system described in chapter 2 was able to perform an efficient coupling control between wedge and press connector by setting the swivel angle to 0° and detecting the echo from the inner borehole of the connector, it turned out that this is no longer possible with the new conventional approach with a fixed incidence angle of 26°. In this case the echoes from the inner borehole are too weak and cannot be evaluated. Instead, the interface echo from the wedge/connector transition can be used because its amplitude depends on the existence (or non-existence) of the coupling. In order to detect this echo effectively, it was necessary to modify the geometry of the wedge so that the interface echo is reflected by an appropriate surface and back-propagated to the transducer. By using this principle, the pulse-echo concept of the original prototype could be kept and an alternative pitch catch arrangement with additional hardware effort could be avoided.

### *3.3 Mode of operation*

The mode of operation of the new inspection system is displayed in Fig. 6 by a 2-D numerical EFIT simulation [3] of wave propagation in one of the six wedges of the gripper and in the press connector, respectively. The snapshots represent the absolute value of the particle velocity vector in equidistant time intervals of 0.5 μs. They demonstrate how the initial P wave front enters the wedge under an incidence angle of 26°, how it is refracted at the wedge/connector interface under an emergent angle of 60°, how it is reflected by the crack in the upper left corner of the connector and back-propagated to the pulse-echo transducer. Moreover, it can be seen from the pictures that the P-P reflection from the interface is reflected by an additional reflecting surface in the upper left part of the wedge and is also back-propagated to the transducer. This latter echo is used for the coupling control of the inspection gripper, preventing unsound inspections in the case of imperfect coupling conditions.

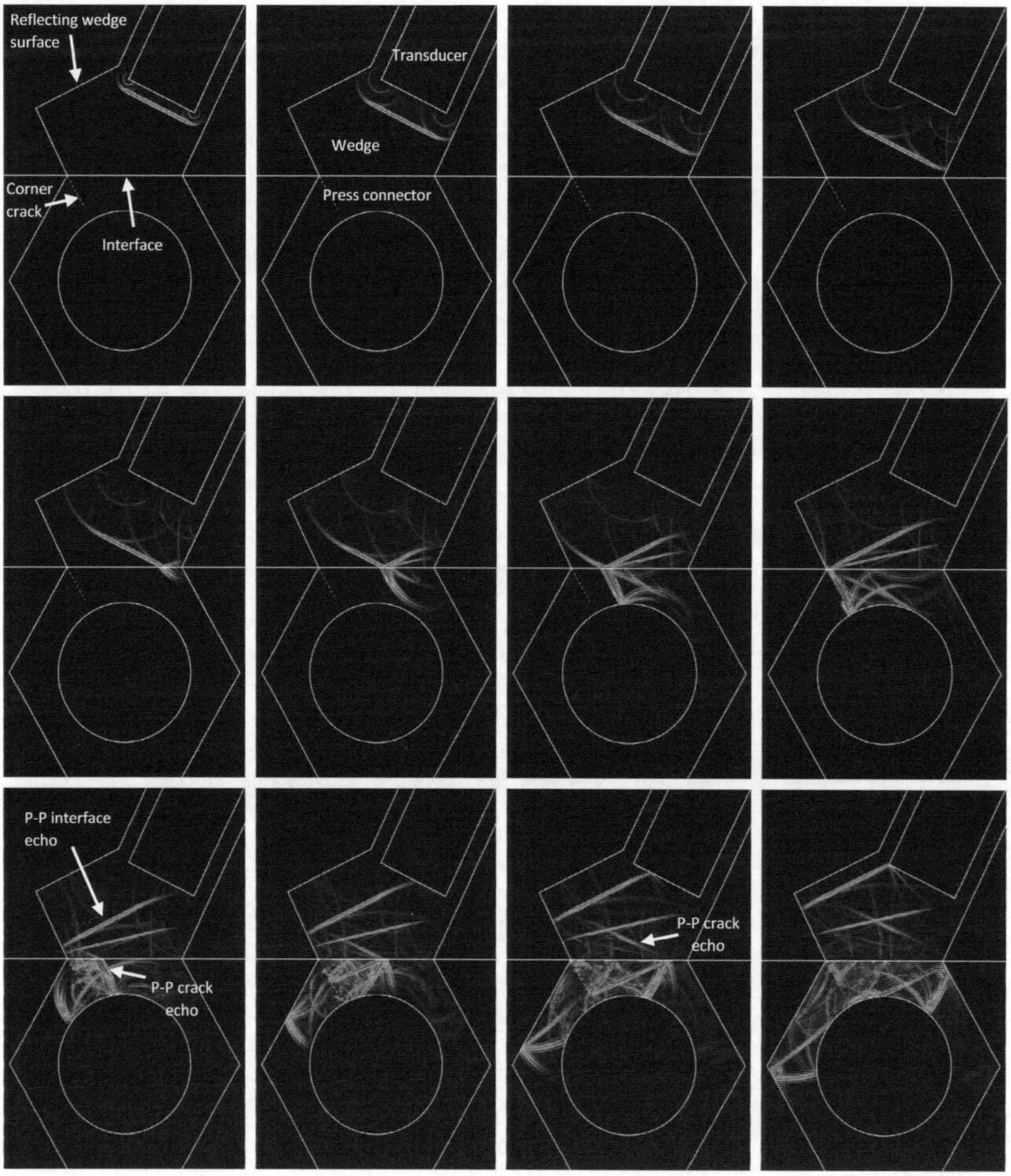


Figure 6. Numerical EFIT simulation of ultrasonic wave propagation in the wedge and the press connector caused by a 26° incident P wave field at 10 MHz centre frequency. Relevant interactions are at the interface (horizontal line), the crack in the press connector (upper left corner), and the reflecting surface of the wedge (upper left part).

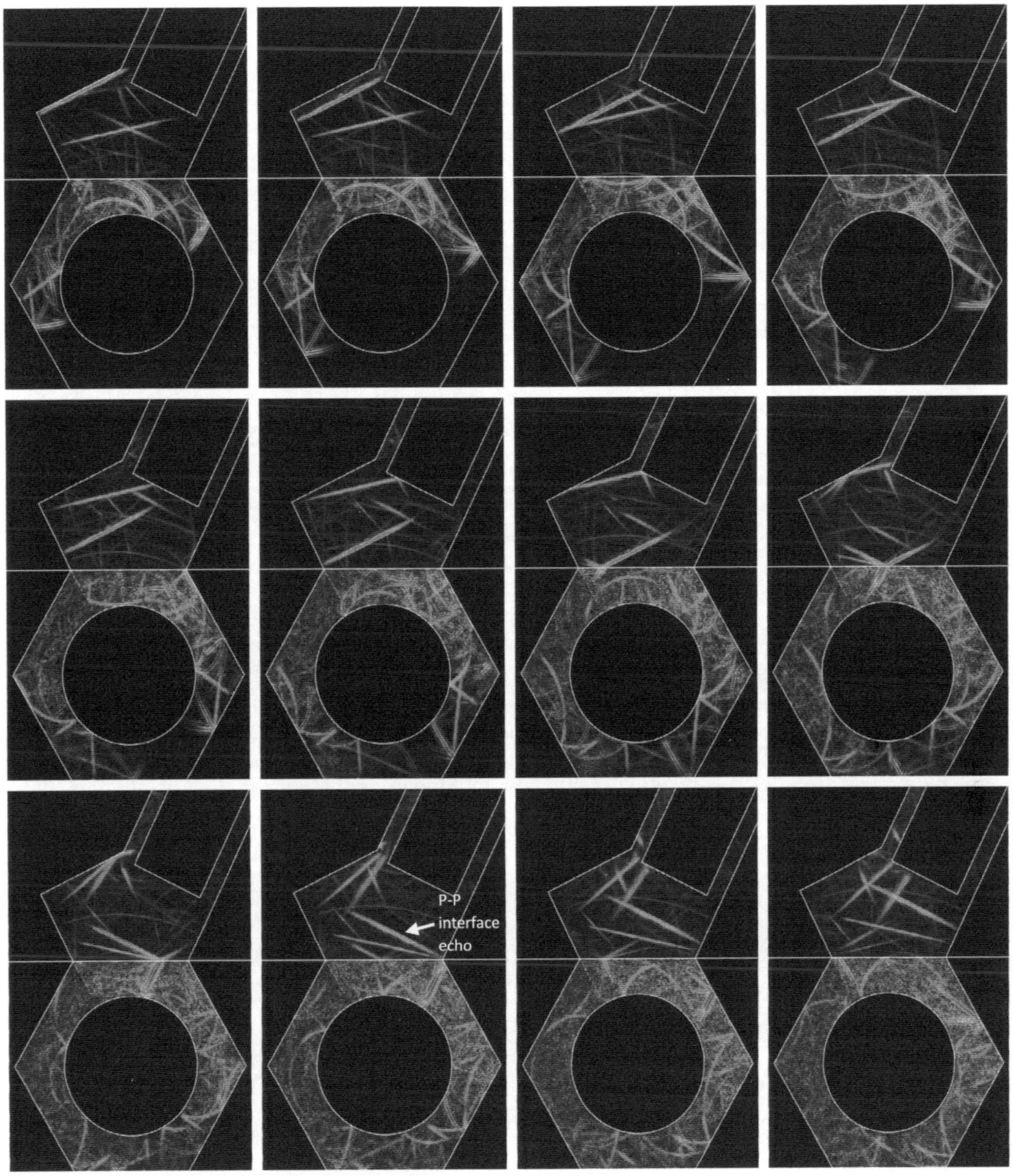


Figure 6 (continuation). Numerical EFIT simulation of ultrasonic wave propagation in the wedge and the press connector caused by a 26° incident P wave field at 10 MHz centre frequency. Relevant interactions are at the interface (horizontal line), the crack in the press connector (upper left corner), and the reflecting surface of the wedge (upper left part).

In Fig. 7 three calculated pulse-echo signals detected at the transducer aperture are shown. They demonstrate that both the appearance of a crack and a proper coupling based on the interface echo can be identified. However, the difference between the interface echoes in cases of perfect acoustic coupling and total decoupling, respectively, is rather small (Fig. 7, on the right).

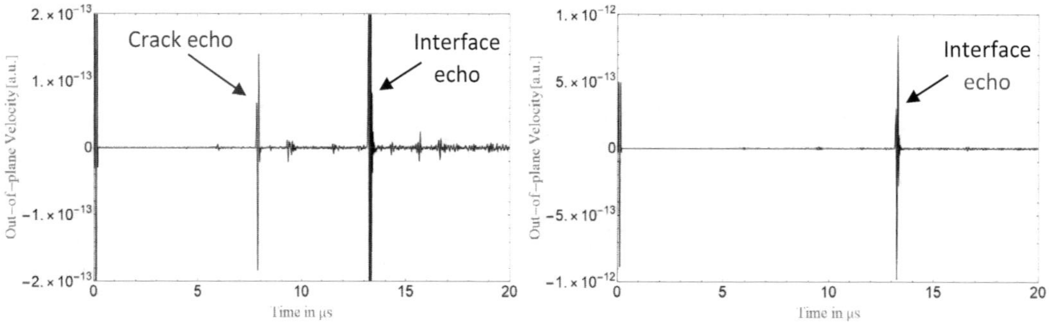


Figure 7. Calculated pulse-echo signals from the model shown in Fig. 6. Black curve: full acoustic coupling without crack; blue curve: full acoustic coupling with crack; red curve: no acoustic coupling.

### *3.4 Implementation*

Based on the concept approved by the numerical simulations, appropriate wedges with additional reflecting surface were designed and manufactured. One of them is shown in Fig. 8 together with a standard normal transducer and a press connector with a reference notch in the upper right corner. Six of these wedges and their corresponding transducers are integrated in the new inspection gripper which is depicted in Fig. 9. It is controlled by a PCUS® *pro* Single ultrasound electronics together with an appropriate multiplexer for the switching of inspection surfaces.

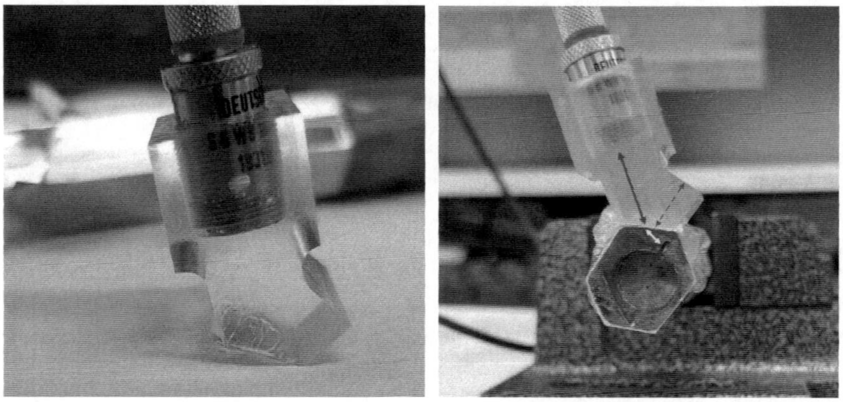

Figure 8. Transducer set-up with adapted wedge incorporating an additional reflecting surface for evaluation of the interface echo.

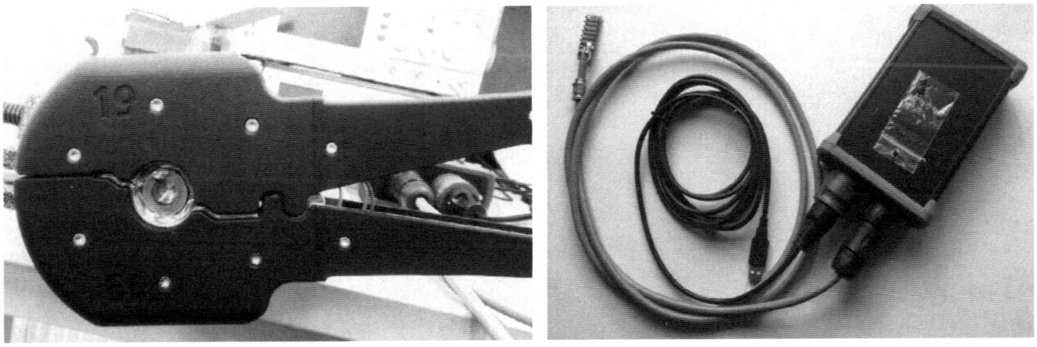

Figure 9. New inspection gripper with six integrated standard normal transducers (on the left) and the proprietary ultrasound electronics PCUS® *pro* Single (on the right).

### *3.5 Preliminary results*

Based on the new wedge configuration displayed in Fig. 8, first measurements on real press connectors with artificial defects were performed in order to demonstrate the ability to detect cracks and to monitor adequate coupling conditions at the interface. The results of these preliminary inspections are shown in Figs. 10 and 11. It is worth mentioning that the difference between the interface echo amplitudes in cases of perfect coupling (Fig. 10, on the left) and complete decoupling (Fig. 10, on the right) is much larger in the experiment than in the 2-D simulation (Fig. 7, on the right). This fact is still under clarification but improves the stability and reliability of the coupling control significantly.

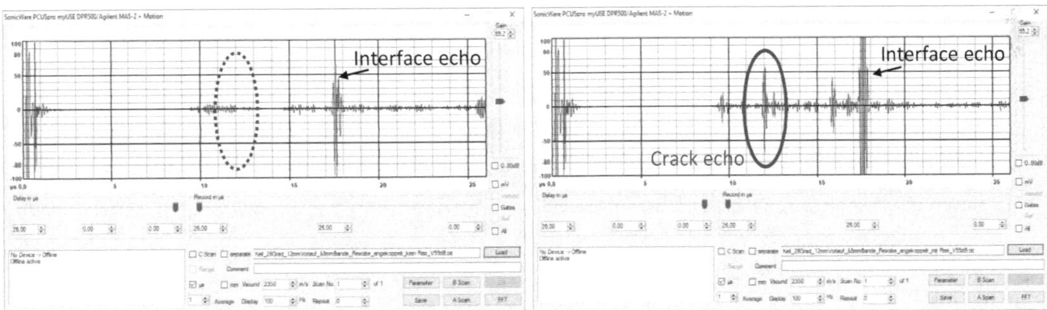


Figure 10. Results of preliminary measurements with the new inspection gripper. On the left: connector without crack (full coupling); on the right: connector with corner crack (marked in red).

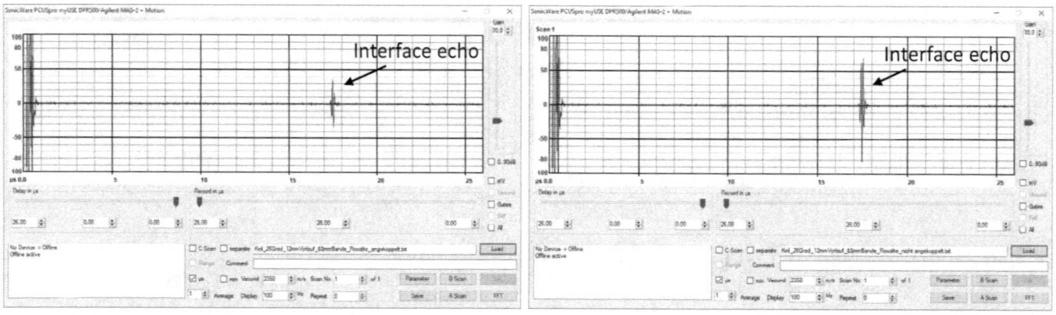


Figure 11. Results of preliminary measurements with the new inspection gripper. On the left: full acoustic coupling at the interface (no crack); on the right: complete decoupling at the interface yielding a 6 dB increase of the interface echo.

## 4. Conclusions

A robust ultrasonic inspection gripper for crack detection in press connectors of railway catenary wires was successfully developed and validated. The prototype is currently tested in the field. It represents the final result of a simulation-based design, a long-lasting development, and a careful physical, technical, and economic consideration of the pros and cons of phased array and conventional single-channel UT systems. In the present case the conventional approach with six multiplexed normal transducers proved superior to the more complex phased array solution. With the new system it is possible for the first time to quickly detect the formation of cracks at an early stage and to identify and replace potentially damaged press connectors accordingly. The presented technical solution is protected by an utility model and by a European patent.

**Acknowledgements**

We would like to thank Dipl.-Ing. Harald Rüdiger of I-VIS GmbH Luzern, Switzerland, for continuous support during the development. We also thank Dr. Peter Krüger from Fraunhofer IKTS for providing the X-ray CT shown in Fig. 1.

**References**

1. IEC 61238-1-1:2018, Compression and mechanical connectors for power cables - Part 1-1: Test methods and requirements for compression and mechanical connectors for power cables for rated voltages up to 1 kV (Um = 1,2 kV) tested on non-insulated conductors, International Electrotechnical Commission.
2. C Richter, H Funke, F Schubert and H Heuer, 'A new compact ultrasound electronics for high-frequency NDE applications between 5 and 150 MHz', In: Proc. of International Congress on Acoustics (ICU), Bruges, Belgium, Sept 3-6, 2019.
3. F Schubert, 'Numerical time-domain modeling of linear and nonlinear ultrasonic wave propagation using finite integration techniques – Theory and applications', Ultrasonics 42, 221-229, 2004.

# Condition Monitoring of Toroidal Roller Bearings Using Acoustic Emission Technique

Housam MOHAMMAD [1], Frantisek VLASIC [1], Pavel MAZAL [1]

[1] Brno University of Technology, Faculty of Mechanical Engineering, Institute of Machine and Industrial Design
Technicka 2, Brno 61669, The Czech Republic;

e-mails: housam.mohammad@vutbr.cz, frantisek.vlasic@vutbr.cz, pavel.mazal@vut.cz

**Abstract:** In this paper the results of the tests on a newly produced type of toroidal bearings were demonstrated. Acoustic Emission (AE) was used as a non-destructive testing technique. The aim was to find the characteristics of the Rolling Contact Fatigue (RCF) in these special bearings during testing, and to find a relationship be-tween the temperature of the tested bearing with its other parameters. The main concentration of the tests has been on the period before the appearance of pitting. It is thought that by analysing and studying the Acoustic Emission (AE) signal in the period before pitting starts to take place, and by extracting temperature of the bearing parts and trying to find a relationship to the temperature it is possible to specify the pattern of the signal which will distinguish the initiation of pitting, which will enable us later to better understand the behaviour of the specimen during the last stages of the test and to expect to some degree the initiation of pitting. The experiments took place in a specially-designed stand for this type of bearing to obtain the best results from AE sensors. This study is an initial step towards other more advanced steps to better understand the crack initiation and pitting mechanisms in toroidal bearings, and then to extend it to self-align bearings.

**Keywords:** Acoustic emission, rolling contact fatigue, toroidal roller bearings.

## 1. Introduction

Toroidal roller bearings are single row bearings with long, crowned rollers. The con-cave raceways in the inner and outer rings are concentric relative to the centre of the bearing. The raceway profiles are matched to each other and ensure optimum distribution of stresses in the bearing as well as low operating friction. [2]
The rollers are self-guiding. They will always automatically adopt the position at which the load is distributed over the length of the roller, Fig. 1

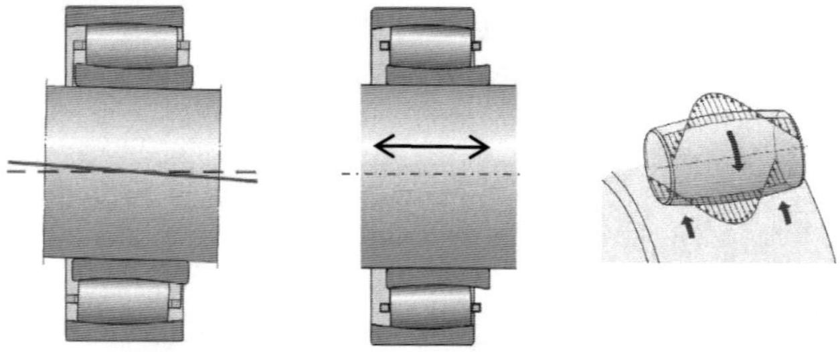

Accommodation of angular misalignment | Accommodation of axial misalignment | Optimal distribution of stresses

**Fig. 1**. Basic characteristics of the toroidal bearing [1]

Where a shaft is liable to temperature-induced elongation and misalignment defects, the non-locating bearing is a particularly important concept. In this case, toroidal roller bearings have proved ideal as non-locating bearings, Fig. 2.

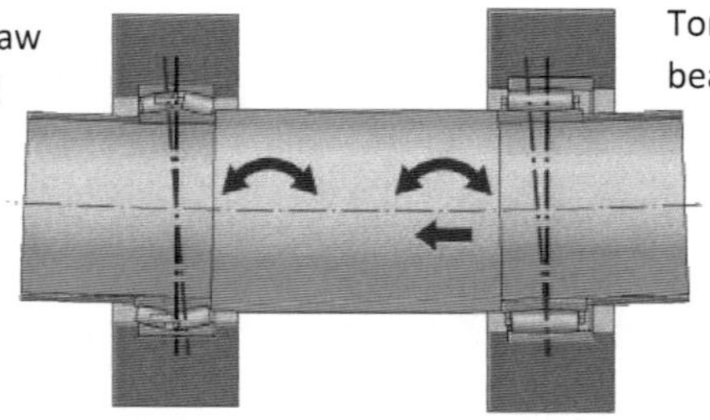


**Fig. 2**. The non-locating concept [2]

In comparison with normal non-locating bearing arrangements, toroidal bearings offer significant advantages: Substantial changes in shaft length are compensated without constraint between the raceways and the rolling elements within the bearing. The design work involved in other solutions is unnecessary. Even more considerable axial displacements have no effect on the locating bearing. There is no axial distortion of the bearing system [2].

The Acoustic Emission method (AE) is a Non-destructive Testing (NDT) method that depends on capturing and processing acoustic waves, which are caused by local changes in the material. Acoustic waves propagate from the place of origin of the defect to the surface of the material, so we can locate the defects and also monitor the course of their origin [5].

Contact fatigue is a very unfavorable phenomenon and it is better to minimize it as much as possible. Rolling Contact Fatigue (RCF) occurs in surface layers by repeated contact of two non-conformal surfaces [6]. Contact bodies are contacted by mutual contact at a point, on a straight line or along a curve. Because the surfaces of the bodies meet in a small area, very high local contact pressures occur. In pure rolling, the main stresses are compressive. Shear stresses reach their maximum at a certain depth below the surface. In these places, cracks arise and grow, which penetrate the surface and lead

to peeling or crumbling of small volumes of material. This creates small cavities, which are collectively referred to as pitting. The main parameters influencing the contact fatigue strength include Hertz stress, number of cycles, sur-face quality and hardness, temperature, and presence of lubricants [3].
The acoustic emission AE method is used to precisely capture the beginning of pit-ting. It is one of the methods of non-destructive material testing based on the principle of elastic stress wave propagation in the material [4].
Combining the AE monitoring and the AE source locator can be used to detect the incipient damage and to forecast the position of the damage in rollers, and this technique allows the monitoring of the deterioration rate of the rolling elements [12]
The basic parameters of the acoustic emission signal evaluation include Root Mean Square RMS (RMS value), counts, and frequency characteristics of the signal [5]. The most commonly used time-domain acoustic emission signal parameter is RMS. A very often used parameter of the time domain signal is the counts parameter, which is defined as the number of overshoots over the set threshold level per unit of time [6].
A wide range of studies has been conducted to study the rolling contact fatigue in ball bearings and spherical bearings, in the case of toroidal roller bearings there is almost none. Toroidal bearings share some similarities with spherical roller bearings, except the fact that they can also accommodate axial displacement as well as angular misalignment. In this paper the AE method has been used to experimentally study the behavior of toroidal bearing and the beginning of pitting under various loading and rotational regimes.

## 2. Material and equipment

The experiments took place in one of the facilities of the Czech bearing company ZKL, which is the company that produces the toroidal bearing under test, and where the special testing stands were designed and produced as well. As shown in the figure 2, the specimen, which is a toroidal bearing of the type C4030V, the material of all bearing parts is bearing steel 100Cr6 which is hardened and tempered to hardness HRC 58 for the rings and HRC 61 for the rollers. The basic geometrical specifications of the tested bearings are: Bore diameter (d)= 150mm, Outer diameter (D)= 225mm, Width (B)= 75mm, Weight= 10.50 kg. The radial internal tolerance of the bearing is 150 μm. The tested bearing has been put under variable controllable loads and it runs according to various testing regimes, as in table 1.

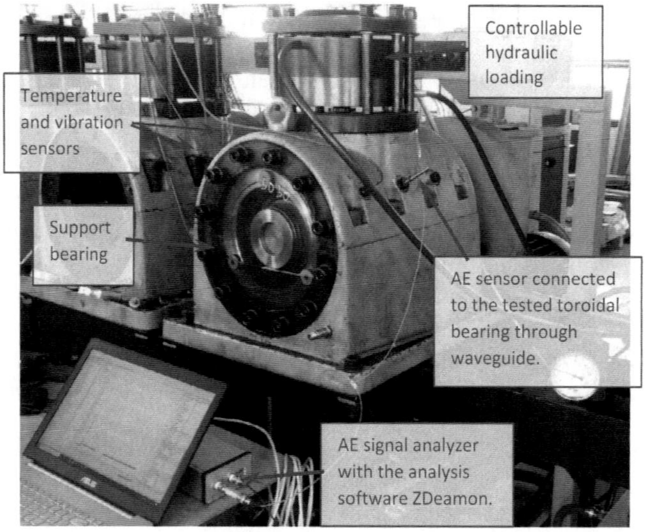


**Fig. 3**. The setup of the toroidal bearing testing station

The main part of the stand is the housing that contains the tested bearing with two support bearings (cylindrical bearings), and the hydraulic loading system is mounted on this housing, as shown in fig. 3. The upper part of the split housing contains two waveguides, which are responsible for transmitting the AE signal from the surface of the outer ring of tested bearing to outside the housing where it can be picked up using AE sensors, as shown in the schematic drawing in fig. 4. For better accuracy, two AE sensors were attached to the two waveguides, and the results were obtained from both of them. The AE sensors used in this experiment are from the type IDK-14 Built-in preamplifier 20 dB, frequency range [10-400 kHz] and the AE monitoring system used is the analyzer DAKEL-ZEDO with two channels. The sampling frequency used for frequency spectrum computation is 5 MHz. The frequency bandwidth was limited by digital filters: High-pass filter 50 kHz and Low-pass filter 400 kHz (software switchable). The global measurement period for AE parameters was set to 1 second, because it was anticipated that the test will run for many days.

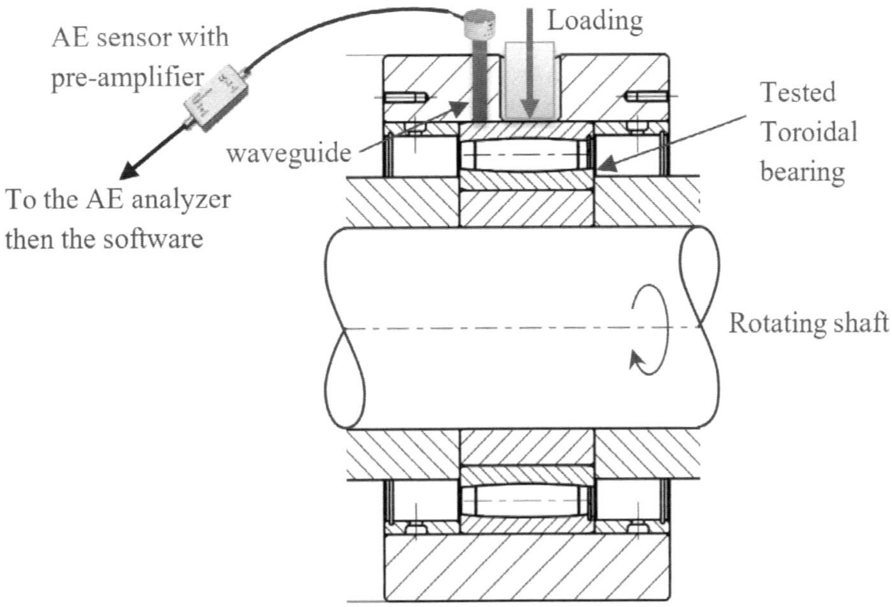


**Fig. 4.** Schematic drawing of the main part of the experiment

**Results and discussion**

The bearing was loaded by different loads during three rotation speeds. Two AE sensors were used to acquire the signals during the operation, they were fixed on the end of the signal waveguides that are in touch with the outer ring of the tested bearing. The received signal was processed and analyzed using DAKEL Zedo analyzer. Three parameters of AE were used to process and analyze the signals, they are: RMS, Energy, and Count of hits. The processed signals were divided to three stages according to the rotation speed of the bearing.

**Table 1.** The three stages of rotation speeds and loading.

| | Stage 1 | | | Stage 2 | | | Stage 3 | | |
|---|---|---|---|---|---|---|---|---|---|
| Speed (rpm) | 250 | | | 500 | | | 750 | | |
| Loading (MPa) | 2.4 | 3 | 3.6 | 2.4 | 3 | 3.6 | 2.4 | 3 | 3.6 |
| Duration (hour) | 1 | 1 | 1 | 0.5 | 1 | 1 | 0.5 | 1 | 1 |

From the experiment, it was noticed that:

In stage1, where the rotation speed is relatively slow (250 rpm), the effect of different loading was significantly small. The RMS value of the signal was very low, less than 0,1 mV. Although there was a small peak immediately after the loading with 3,6 MPa, which is related to the running-in phase of the bearing. The Energy parameter was almost redundant at this stage. And the count hit just followed the same pattern.

Stage 2, in this stage the rotation speed was raised to (500 rpm) which was reflected in a big impact on the values of RMS of the signal that reached the value of 0,3 mV. By implementing the same previous loading regime on the bearing we notice a very small, almost negligible effect of change when raising the loading during the same rpm. The two other parameters, Energy and count of hits, as the RMS, showed a relevant change in values according to the increase of rotation speed, but almost no change for the increase in loading.

Stage 3, where the rotation speed reached (750 rpm), we see a respectively different relationship pattern between the loading and the signal. In this stage the increase of loading had a proportionate impact on the RMS values and the other two parameters, as shown in the Fig. 5.

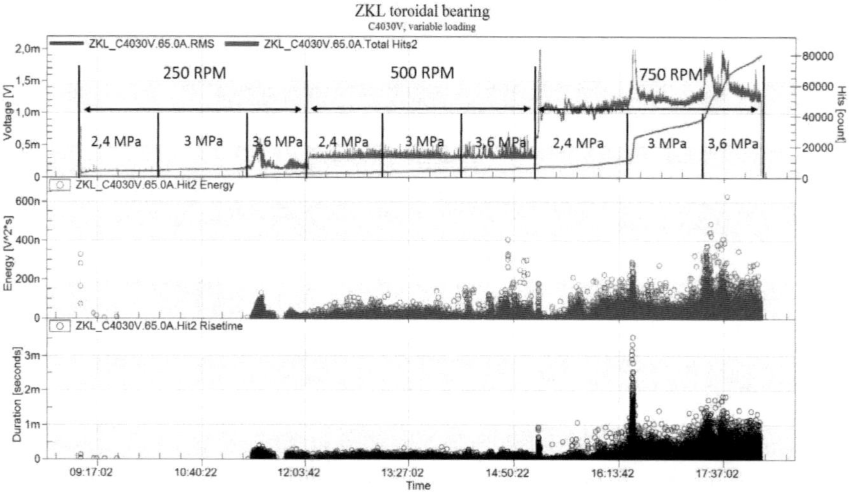


**Fig. 5**. The relationship between AE parameters, the rotation speed, and loading of the toroidal bearing.

It is noticed from the figure that the increase of the rotation speed of the bearing has a bigger effect on the signal parameters. The increase of loading of the bearing during the relatively small rotation speeds has a very low effect on the signal parameters, as in fig. 5, the energy of the signal (the single hit energy which equals the surface under the hit curve) and the duration of the hit. While the same change in loading during higher rotation speed has resulted in a more proportionate increase of the values of signal parameters. These notes will be taken into consideration while choosing the conditions for the following series of tests that are planned on this type of bearing.

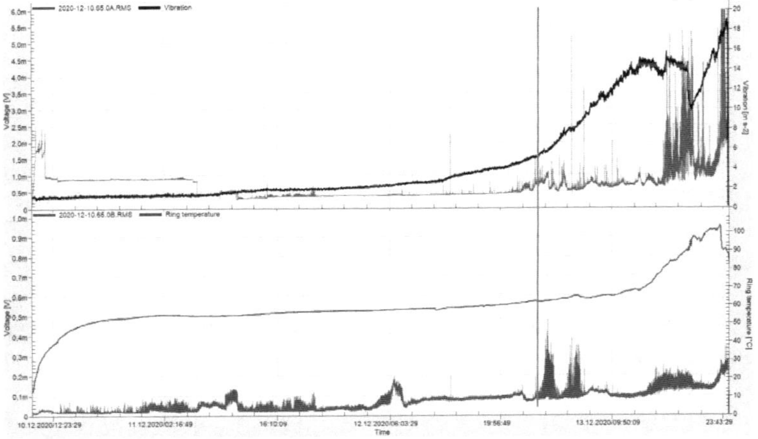


**Fig. 6.** RMS of the AE signal from the two AE sensors, the upper from the bearing outer ring, and the lower from the housing, with the values of vibration (RMS) on the upper graph, and temperature of the outer ring on the lower.

In the next experiment, the rotation speed was fixed at 375 rpm, and the loading was also fixed at 172,5 kN. Two AE sensors were used, the first (A) was attached to the outer ring of the tested bearing, with the help of a waveguide, the second (B) was attached to the housing of the bearing to act as a reference. Then a temperature sensor and vibration sensor were added to the system, both of them attached to the outer ring of the bearing. The results of the test, which lasted for about 3.5 days, were presented in figure 6. The upper graph in fig. 6 shows the RMS of the AE signal received from the sensor A, plotted with the vibration curve, which gives the RMS of the vibration value, and the lower graph shows the RMS of AE signal received from the sensor B, plotted with the temperature curve.

As shown from the upper graph in fig. 6, at the beginning of the test there were high level of AE due to the running in stage of the new tested bearing, then the signal stabilized for most of the time of the test until the third day when the beginning of the assumed effects of RCF started to take place. This assumption comes from the big increase in the energy, and RMSAE of the AE hits as well as the increase in vibration level which also indicated the initiation of the cracks that developed later to form the pits, which increased in number to raise the value of the vibration dramatically and with it the temperature of the bearing, and to trigger the stop of the test.

The main observation of this graph is that the temperature of the outer ring, and supposedly of the inner ring as well, started to rise after long time of the start of increase in vibration level which indicated the beginning of pitting formation on the tracks in the inner and outer rings. This is important because it means that the temperature of the rings is indicative but less responsive to the changes regarding the effects of RCF in the bearing parts. In other words, the temperature of the inner and outer rings is not good parameter for fast and accurate representation of the changes in the bearing parts regarding the RCF effects.


ORGANISING SECRETARIAT: GUARANT International spol. s r.o. European NDT&CM Days 2021:
Českomoravská 2510/19, Praha 9 - Libeň, Česká republika
E-mail: endtcm21@guarant.cz Tel: +420 284 001 444, Fax: +420 284 001 448

## Conclusion

This paper demonstrated the results of testing a newly produced type of toroidal roller bearings using AE as a NDT method. The received signal was analyzed and divided into three stages according to the rotation speed, and the same gradual increase of loading regime was applied on each stage. The AE signal parameters were more sensitive to the increase in rotation speed than to the changes of loading, so it was concluded that the selection of the rpm would be more important in the following tests.
It was also found that the temperature of the bearing rings is not a reliable parameter for fast and accurate detection of the beginning of the RCF effects in the bearing, but it still can be used as an indicative parameter.
This work is the first step of a long-term research to study and monitor toroidal bearing behavior with regard to the rolling contact fatigue, and it is still in its first stage, where we are collecting, analyzing and presenting the data.

## Acknowledgment

This work was supported by the project of the Technology Agency of the Czech Republic No. TH 02010306 " Research and development of new generation design and technology of spherical roller thrust bearings".

## References

[1] SKF CARB® toroidal roller bearings, a revolutionary concept. © Copyright SKF 2004, Publication 5102 E. January 2004.
[2] Schaeffler Technologies AG & Co. KG, www.schaeffler.de/en. © Schaeffler Technologies AG & Co. KG, Issued: 2015, April.
[3] HARRIS, Tedric A. Advanced Concepts of Bearing Technology [online]. CRC Press, 2006. DOI: 10.1201/9781420006582. ISBN 9781420006582.
[4] KOPEC, Bernard. Nedestruktivní zkoušení materiálů a konstrukcí: (nauka o materiálu IV). Brno: Akademické nakladatelství CERM, 2008. ISBN 978-80-7204-591-4.
[5] LYMPERTOS, Efstratios M. a Evangelos S. DERMATAS. Acoustic emission source location in dispersive media. Signal Processing [online]. 2007, 87(12), 3218-3225.
[6] NOHÁL, L, F HORT, J DVOŘÁČEK a P MAZAL. An experimental investigation of rolling contact fatigue of steels using acoustic emission method. Insight - Non-Destructive Testing and Condition Monitoring [online]. 2013, 55(12), 665-669.
[7] GHALAMCHI, Behnam, Jussi SOPANEN a Aki MIKKOLA. Modeling and Dynamic Analysis of Spherical Roller Bearing with Localized Defects: Analytical Formulation to Calculate Defect Depth and Stiffness. Shock and Vibration [online]. 2016, 2016, 1-11
[8] JOHNSON, K. L. Contact mechanics. New York: Cambridge University Press, 1985. ISBN 0-521-25576-7.
[9] CHOUDHURY, A. a N. TANDON. Application of acoustic emission technique for the detection of defects in rolling element bearings. Tribology International [online]. 2000, 33(1), 39-45.

[10] MBA, D. Acoustic Emissions and Monitoring Bearing Health. Tribology Transactions [online]. 2003, 46(3), 447-451.
[11] SCHWACH, D a Y GUO. A fundamental study on the impact of surface integrity by hard turning on rolling contact fatigue. International Journal of Fatigue [online]. 2006, 28(12), 1838-1844.
[12] RAHMAN, Md. Ziaur, Hiroaki OHBA, Takashi YAMAMOTO a Takeo YOSHIOKA. A Study on Incipient Damage Monitoring in Rolling Contact Fatigue Process Using Acoustic Emission. Tribology Transactions [online]. 2008, 51(5), 543-551.
[13] HASE, Alan, Hiroshi MISHINA a Masaki WADA. Correlation between features of acoustic emission signals and mechanical wear mechanisms. Wear [online]. 2012, 292-293, 144-150.
[14] PANDYA, D.H., S.H. UPADHYAY a S.P. HARSHA. Fault diagnosis of rolling element bearing with intrinsic mode function of acoustic emission data using APF-KNN. Expert Systems with Applications [online]. 2013, 40(10), 4137-4145.
[15] FERRANDO CHACON, Juan Luis, Vassilios KAPPATOS, Wamadeva BALACHANDRAN a Tat-Hean GAN. A novel approach for incipient defect detection in rolling bearings using acoustic emission technique. Applied Acoustics [online]. 2015, 89, 88-100.
[16] VAN HECKE, Brandon, Jae YOON a David HE. Low speed bearing fault diagnosis using acoustic emission sensors. Applied Acoustics [online]. 2016, 105, 35-44.
[17] COCKERILL, A, A CLARKE, R PULLIN, T BRADSHAW, P COLE a KM HOLFORD. Determination of rolling element bearing condition via acoustic emission. Proceedings of the Institution of Mechanical Engineers, Part J: Journal of Engineering Tribology [online]. 2016, 230(11), 1377-1388.
[18] SHARMA, Ram Bihari a Anand PAREY. Modelling of acoustic emission generated in rolling element bearing. Applied Acoustics [online]. 2019, 144, 96-112.
[19] MAHMOUD, Houssam, Pavel MAZAL a Frantisek VLASIC. Relationship between acoustic emission signal and loads on pneumatic cylinders. Nondestructive Testing and Evaluation
[20] SCHNABEL, S., P. MARKLUND, R. LARSSON a S. GOLLING. The detection of plastic deformation in rolling element bearings by acoustic emission. Tribology International [online]. 2017, 110, 209-215.

# Vibration-Based Feature Extraction for Artificial Damages in Small Scale Cutting Discs of Tunnel Boring Machines

Sebastian PRIEBE[1*], Lukas BRACKMANN[2], Arne RÖTTGER[3], Günther MESCHKE[1] and Inka MUELLER[4]

[1*]Institute for Structural Mechanics, Ruhr University Bochum, Universitätsstr. 150, Bochum, 44801, NRW, Germany.
[2]Chair for Materials Technology, Ruhr University Bochum, Universitätsstr. 150, Bochum, 44801, NRW, Germany.
[3]Chair of New Manufacturing Technologies and Materials, University of Wuppertal, Bahnhofstr. 15, Solingen, 42651, NRW, Germany.
[4]Bochum University of Applied Sciences, Am Hochschulcampus 1, Bochum, 44801, NRW, Germany.

*Corresponding author(s). E-mail(s): Sebastian.Priebe@ruhr-uni-bochum.de;

**Abstract**

The efficiency of tunnel boring machines (TBM) depends on the condition of the cutting tools used, whereby the condition of cutting discs is hardly monitored. As cutting discs build a rotating system, an approach of vibration-based feature extraction to monitor cutting discs is proposed. The feature extraction via Hilbert transformation is refined and applied to experimental data to detect artificial damages. Tests are performed with the RUB-Tunneling-Device, which is a model of the cutting system of TBMs. Here, it is used to test small scale cutting discs with damages of different sizes. A set of 65 experiments with multiple discs of the same design and increasing damage is presented. The extracted features show a correlation between damage size and damage indicator. The variation of damage indicators among experiments of one disc is smaller than the variation among experiments with different discs. However, with sufficient data, it is possible to identify the damaged state and separate it from the baseline.

**Keywords:** condition monitoring, structural health monitoring, vibration based feature extraction, tunnel boring machines, cutting discs

# 1 Introduction

The usage of vibration-based condition monitoring (CM) for expensive rotating machinery with stationary operational and fixed environmental conditions, like turbines, pumps, gearboxes and tooling machinery, is state of the art. Its goal is to identify upcoming damages before component failure to enable condition-based maintenance. In this study, an approach for CM with vibration-based methods is proposed for applications where parts are exposed to wear, rough environmental and non static operational conditions, such as cutting discs in tunnel boring machines (TBMs). Aiming for an improved efficiency in operation and determination of optimal set-up times with simultaneously low downtimes, the necessity of examining the wear condition using condition monitoring methods of the cutting tools involved is gaining importance. A great number of cutting discs are located in the front shield of a TBM suited for hard rock conditions. During operation, these cutting discs are rotating on the rock face that is to be mined. Sufficient forces applied by the cutting discs cause cracking of the rock and thus rock excavation by chipping. These forces typically reach around 250 kN per disc, but can exceed this value by a factor of four during operation [1]. This force combined with impact loading can cause initial damages located at the edge of the disc, which this work will focus on.

Different methods to use vibration-based data for monitoring exist. One method is operational modal analysis (OMA) which uses natural frequencies, modal masses, modal shapes and modal damping as parameters to identify damages. Another method uses more general observations of the data such as peak-to-peak values, root mean square values and statistical moments for CM. OMA and condition monitoring are mainly used for load bearing structures exposed to changing environmental and operational conditions (EOC) and stationary, rotating machines used at mostly constant EOC, respectively. This study investigates the possibility to monitor cutting discs of TBMs with vibration-based methods used in condition monitoring. These discs can be considered as rotating machinery, although EOC will differ greatly. An extensive overview about vibration-based methods is given by Avci et al. in [2]. They cover over 100 studies covering parametric, non-parametric, machine learning and neural network studies. Recent research using modal parameters in [3] presents the comparison of two approaches to detect a delamination in fiber reinforced polymer structures. Lab-based approaches using a shaker and an IR-camera are compared with PWAS and surface temperature sensors. Latter makes it possible to achieve permanent monitoring for aviation purposes. Furthermore, Entezami and Shariatmadar, in [4], are improving the ensemble empirical mode decomposition method to deal with residual noise and redundant modes. Also of interest to this work on the topic of condition monitoring is the work of Caldero and Zoeke [5], who analyse features in the frequency domain while using wireless sensors to be able to attach sensors directly to moving parts which are mostly the reason for necessary maintenance. Some of the latest methods and technologies appropriate for SHM and CM, with special interest to gearboxes and wind blade monitoring, are given by Antoniadou et al. in [6]. In addition, they use the empirical mode decomposition in combination with an AM-FM demodulation to show that the power of damage sensitive intrinsic mode functions can be used to detect damaged teeth in spur gears.

The goal of this study is to show that small damages in cutting discs can be detected long before the risk of catastrophic disc failure and therefore potential damage to other parts occur. Similar to gearboxes, TBM discs can either fail due to fatigue or due to load peaks leading to spalling. This paper focuses on those failures, which are initiated by a small local change at the cutting edge of the disc. To achieve this, first the experimental setup, named RUB-Tunneling Device (RTD), will be introduced before the feature extraction methodology and the results are presented. This study will then discuss the results and draw the conclusions regarding the possibility of detecting small damages.

# 2 Methods

Experimental data is collected from the RTD on which a load cell and an acceleration sensor are mounted from which data in the form of time-series' with 1 kHz sampling frequency was collected. With the collected data, several steps of post-processing were done before final damage indicators are achieved.

## *2.1 Experimental Setup*

The RTD is a model of the cutting system of a TBM, which is capable of many disc rotations. This makes it possible to observe cutting discs in general for many more rotations than in linear cutting tests. This is realised by using a cylindrical body (counter-body) as surface for the disc to roll on. The cutting discs are made of cold work tool steel (1.2379) and the counter body is made of S235JR. A detailed description of the setup can be found in [7]. This study consists of 65 experiments covering 3 identical cutting discs with up to 4 different damage sizes, ranging from 0.5 to 3.08 mm, per disc. Damages were introduced by hand and increased incrementally. To accurately determine the size of the damages, a high resolution picture of the damage with a scale next to it was taken from which the size was measured (Table 1). The focus of this work is on damage detection of small damages located at the cutting edge of small scale cutting discs. These damages interact with the counter-body only once per revolution. Hence, it is important to have a high number of revolutions in each experiment. An initial force of 1 kN is applied to push the cutting disc against the counter-body, before the RTD is then started with a frequency of 23.4 RPM, which is well within the limits of a real TBM (e.g., [8]). The setup uses a load cell (KM40 - ME-Meßsysteme) and an acceleration sensor (333B30 - PCB Synotech) to measure the normal force acting on the cutting disc, as well as the acceleration in normal direction. While the load cell directly measures the forces acting on the cutting disc, the mounting position needs to be much more specific than that of the acceleration sensor. On the other hand, the acceleration sensor is much more responsive, which can lead to higher noise levels and also picks up vibration caused from different parts of the setup or even the environment (Fig. 1).

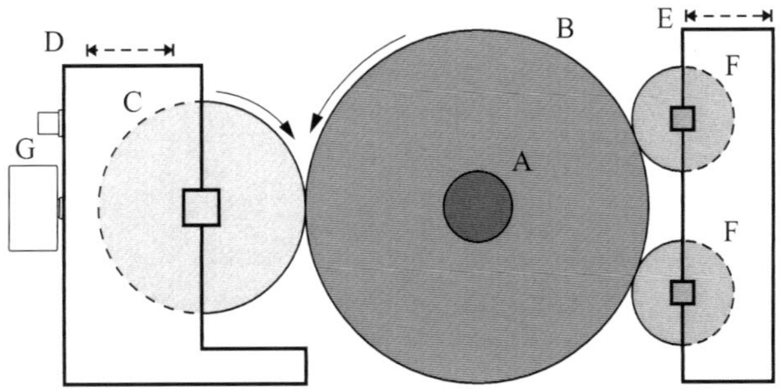


**Figure 1** Experimental setup of the superstructure of the lathe with shaft (A), counter-body (B), cutting disc (C), linear moving carriages (D, E), force balancing rolls (F) and the positions of the two sensors (G). The force sensor is in line with the axis of the cutting disc and the point of contact between cutting disc and counter-body [7].

**Table 1** Summary of experimental dataset

| disc ID | Damage sizes in mm | Number of exp. per damages size |
|---|---|---|
| A (blue) | 0, 0.64, 1.08, 2.09, 2.97 | 8, 6, 6, 6, 6 |
| B (green) | 0, 0.50*, 1.09, 1.98 | 5, 3, 3, 3 |
| C (red) | 0, 1.28, 2.15, 3.08 | 10, 3, 3, 3 |

* Nominal value as not accurately determined.

## *2.2 Data Processing*

To extract features which are sensitive to the presented damages, a number of steps are taken to process the data. These steps, described in detail afterwards, are:

- cutting of the data,
- subtraction of the moving average from the signal,
- generating an envelope function with the Hilbert transform,
- transforming the signal into the frequency domain,
- picking the highest value between 0.45 and 0.55 Hz and
- normalising the data by the total energy of the signal.

Since the measurements are often started before the machine is actually running and is still running after it is stopped, the ends of each measurement are not considered. The data is therefore cut, such that it only contains the quasi-static process of the cutting disc continuously rolling on the counter-body with constant speed. Then, the moving average

$$x_{\mathrm{MA}i} = \frac{1}{2n+1} \sum_{k=i-n}^{i+n} x_k \,, \tag{1}$$

with a boxcar window of $2n + 1 = 21$ samples length is calculated and subtracted from the signal $x_i$,

$$\bar{x}_i = x_i - x_{\mathrm{MA}i} \,. \tag{2}$$

The length of the window is chosen based on the typical time of interaction between damage and counter-body which is observed to be typically 14 to 18 ms. For the first and last $n$ samples this can not be done and are therefore not considered in further processing steps. Using this method removes low frequency components from rotating parts of the RTD, leaving the interaction of the damage and high frequency components behind (Fig. 2). Going on from

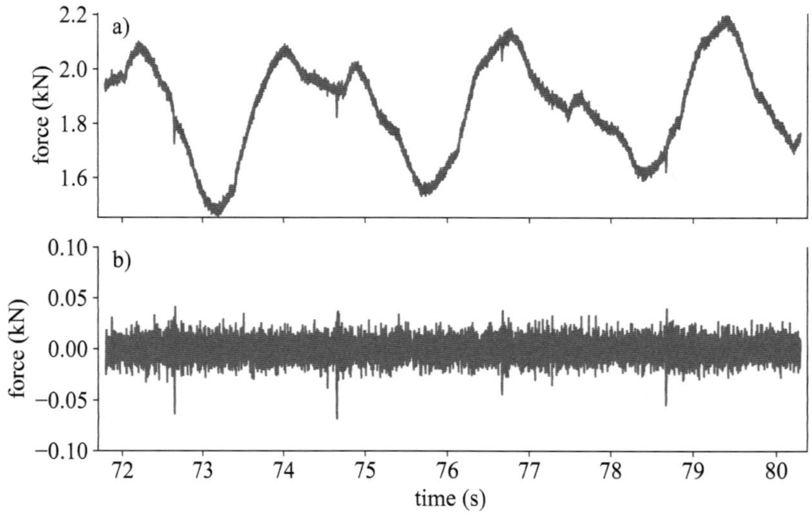


**Figure 2** a) Exemplary time domain data of a disc with a damage interacting every two seconds. b) The same data with the moving average removed from it, still showing the damage interaction.

here, the envelope via Hilbert transform,

$$x_H = F^{-1}(F(\bar{x})2U) = \bar{x} + iH(\bar{x}) \,, \tag{3}$$

is calculated. Here, $F$ is the discrete Fourier transform, $U$ the Heaviside step function an $H$ the Hilbert transform (e.g. [9], implemented in [10]). The now generated envelope signal is then transformed into the frequency domain, where the maximum value in a narrow frequency band (0.45 - 0.55 Hz) around the previously calculated disc revolution frequency is taken as damage feature. The damage frequency $f_\mathrm{d}$ is given by the RTD's RPM and the diameter ratio of counter-body and cutting disc,

$$f_\mathrm{d} = \frac{23,4}{60\ \mathrm{s}} \frac{0.15\ \mathrm{m}}{0.118\ \mathrm{m}} \simeq 0.5\ \mathrm{Hz}.$$

To further improve this feature, it is normalised by the total energy of the signal it is calculated from, since any deviation from a circular surface should cause more vibrations and therefore increase the energy in the recorded signals. Using Parsevals theorem, the total energy $E_\mathrm{T}$ is calculated by

$$E_\mathrm{T} = \frac{1}{N} \sum_{k=0}^{N-1} \|F_k\|^2 , \tag{4}$$

where $N$ is the total number of samples [11]. Finally, damage indicators $DI$ and normalised damage indicators $DI_\mathrm{N}$ can be expressed as

$$DI = \max(F(x_\mathrm{H}) : [0.45, 0.55]) \tag{5}$$

$$DI_\mathrm{N} = \frac{\max(F(x_\mathrm{H}) : [0.45, 0.55])}{E_\mathrm{T}} . \tag{6}$$

# 3 Results

Using the described data of 65 experiments, which cover 3 cutting discs with up to four different damage sizes each, the gathered data, described in (Sec. 2.1) is shown here. The results are presented such, that comparison among discs, sensors and between non-normalised and normalised damage indicators is possible.

The values of damage indicators (DI) are between $10^{-5}$ to $10^{-1}$ for unnormalised data, while normalised DI are between values of $10^0$ and $10^2$. In general, the unnormalised DI from the acceleration sensor are much higher than the DI from the load cell by at least an order of magnitude. All discs show a noticeable trend for unnormalised DI from force data with increasing damage size. Trends from acceleration data are much less incisive for discs A and B. On the other hand, normalised data improves the differentiation between undamaged and damaged DI for disc B, while there is no noticeable change for the other discs (Fig. 3).

When comparing cutting discs among each other, the similarity of the trends of the DI rising with damage size (i.e. the mostly linear gradient), especially for force data, becomes clear. Except for unnormalised DI from acceleration data, disc C consistently shows the highest mean values for all damage sizes and while the variation of disc C is consistently the highest in the undamaged case, it shows the opposite behaviour in all damaged cases (Fig. 4).

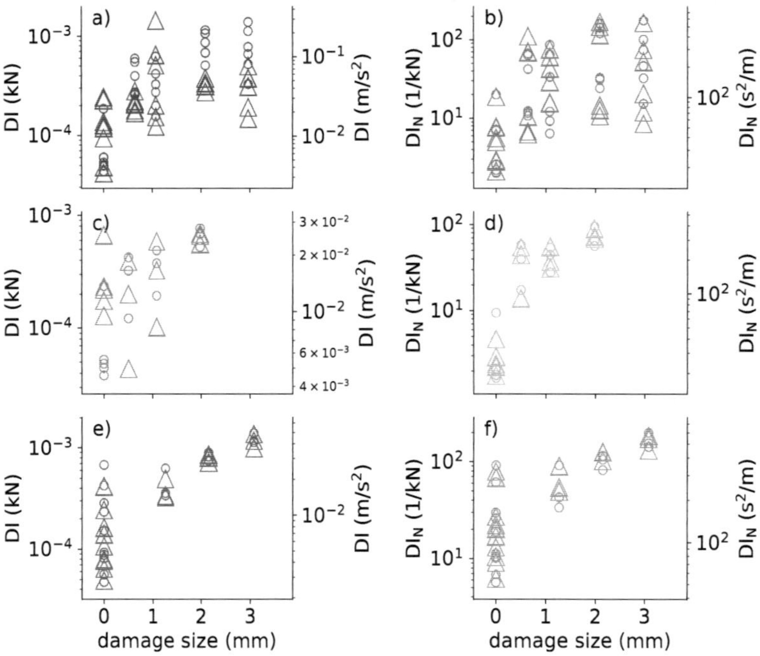


**Figure 3** Normalised and unnormalised damage indicators shown against damage sizes. Disc A shown in blue, disc B in green and disc C in red. DI from force data represented by circles and from acceleration data by triangles. Unnormalised (a) and normalised (b) DI from force and acceleration data for disc A. Unnormalised (c) and normalised (d) DI from force and acceleration data for disc B. Unnormalised (e) and normalised (f) DI from force and acceleration data for disc C.

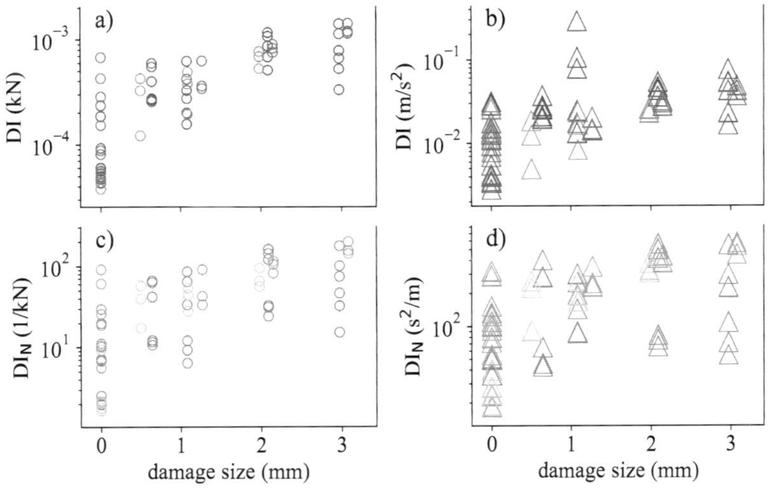


**Figure 4** Normalised and unnormalised damage indicators shown against damage sizes. Disc A shown in blue, disc B in green and disc C in red. DI form force data represented by circles and from acceleration data by triangles. a) Unnormalised DI from force data for all discs. b) unnormalised DI from acceleration data for all discs. c) Normalised DI from force data. d) Normalised DI from acceleration data for all discs.

# 4 Discussion

The data presented in this study clearly suggests a correlation between damage sizes and used DIs. This is shown for different cutting discs as well as different sensors. Although the number of experiments with 65 leads to statistics with small numbers, clear trends for each individual disc are visible.

Using Parsevals theorem to normalise the data yielded mixed results. The variability of the data does decrease in some and increase in other cases. For example, the normalisation for damage sizes 0.64 and 2.09 mm of disc A greatly decreases the variability, but increases it for the damage of 1.08 mm size. Another effect of the normalisation is that the DIs from different sensors are no longer separated regarding their values. However, it did not always improve the correlation regarding damage size (e.g. (Fig. 3)). This is especially true since the undamaged case shows the same behaviour, suggesting that the damage does not change the total energy of the system in any meaningful way. However, this can only be said for the damage investigated in this study. Other damages, which cause the discs to revolve at inconsistent speeds, would increase the total energy. The presented method would then need to be modified in order to cope with these circumstances, for example by normalising the rotational speed via signal stretching.

The total energy not being a satisfactory means leads to the question of what is causing the variability among the DIs. Since the experiments were done in a workshop, environmental conditions can still be considered negligible. Operational conditions on the other hand may vary greatly, as quite a number of parts needed to be removed from the RTD between experiments with different discs and between experiments in which the damage size was increased. Although measures were taken to observe the effect of changing the setup, no correlation among different variations of dis- and reassembling the setup has been found. Still, changing parts of the machine is likely a source of variability which would not occur during the operation of a TBM. Therefore, it should be viable to focus observations on a single disc until all planned experiments are done in order not to introduce additional variability into the system. Another condition that might introduce a lot of variability in this setup is the application of the initial force. This is done via pressurised air and a pressure gauge. Despite the fact, that the initial force was always set to 1 kN ($\pm 10\%$) the used pressure to achieve this was between 0.5 and 1.5 bar above ambient air pressure. In spite of all the mentioned factors, the DI still show a strong correlation between increasing damage size and increasing value. Future studies will aim for greater consistency of the setup and introduce additional equipment in order to observe other characteristics such as eigenfrequencies of cutting discs as well as the penetration depth.

# 5 Conclusion

A study of 65 experiments with three different discs was presented here. Artificial damages were introduced into the discs and incrementally increased. The used feature extraction based around a moving average filter and the envelope generation via Hilbert transform yielded damage indicators which clearly show a correlation between damage sizes and value of damage indicators. Yet, is has been shown that the system produces a lot of variability despite the fact that some features of the setup can be considered as best case scenarios. Possible sources of the variability have been identified and will be investigated in future studies, which will also cover experiments with counter-bodies cut from rock specimens and operational modal analysis in order to investigate if eigenfrequencies are viable parameters for the monitoring of cutting tools.

# 6 Acknowledgements

The authors gratefully acknowledge the support of the German Science Foundation (DFG) for providing financial support in the framework of projects C7 and C5 of the Collaborative Research Center SFB 837.

# References

[1] Entacher, M., Galler, R.: Development of a disc cutter force and face monitoring system for mechanized tunnelling/Ortsbrustmonitoring und Leistungsprognose bei TBM-Vortrieben. Geomechanics and Tunnelling **6**(6), 725–731 (2013)

[2] Avci, O., Abdeljaber, O., Kiranyaz, S., Hussein, M., Gabbouj, M., Inman, D.J.: A review of vibration-based damage detection in civil structures: From traditional methods to machine learning and deep learning applications. Mechanical Systems and Signal Processing **147**, 107077 (2021)

[3] Bergmayr, T., Kralovec, C., Schagerl, M.: Vibration-based thermal health monitoring for face layer debonding detection in aerospace sandwich structures. Applied Sciences **11**(1), 211 (2021)

[4] Entezami, A., Shariatmadar, H.: Damage localization under ambient excitations and non-stationary vibration signals by a new hybrid algorithm for feature extraction and multivariate distance correlation methods. Structural Health Monitoring **18**(2), 347–375 (2019)

[5] Caldero, P., Zoeke, D.: Multi-channel real-time condition monitoring system based on wideband vibration analysis of motor shafts using SAW RFID tags coupled with sensors. Sensors **19**(24), 5398 (2019). https://doi.org/10.3390/s19245398

[6] Antoniadou, I., Dervilis, N., Papatheou, E., Maguire, A.E., Worden, K.: Aspects of structural health and condition monitoring of offshore wind turbines. Philosophical Transactions of the Royal Society A: Mathematical, Physical and Engineering Sciences **373**(2035), 20140075 (2015). https://doi.org/10.1098/rsta.2014.0075

[7] Priebe, S., Brackmann, L., Alabd-Allah, A., Butt, S., Röttger, A., Meschke, G., Mueller, I.: Comparison of hilbert transform and complex demodulation for defect identification in cutting discs using vibration-based feature extraction. In: European Workshop on Structural Health Monitoring, pp. 564–572 (2020). Springer

[8] Geng, Q., Bruland, A., Macias, F.J.: Analysis on the relationship between layout and consumption of face cutters on hard rock tunnel boring machines (TBMs). Rock Mechanics and Rock Engineering **51**(1), 279–297 (2017)

[9] Cizek, V.: Discrete hilbert transform. IEEE Transactions on Audio and Electroacoustics **18**(4), 340–343 (1970)

[10] The SciPy community: SciPy v1.4.1 Reference Guide. https://docs.scipy.org/doc/scipy/reference/generated/scipy.signal.hilbert.html. [Online; accessed 14-August-2020] (2019)

[11] Oppenheim, A.V., Schafer, R.W.: Zeitdiskrete Signalverarbeitung. R. Oldenbourg Verlag München Wien, München (1999). https://doi.org/10.1515/9783486792966

# Non Destructive Examination Of Cargo Vaporizers And Condensers In Ships Carrying Liquefied Petroleum Gases In Bulk (LPG Carriers)

Tejas INGALE

Injotech India Private Limited, Office No. 44, 5th Floor, Yugay Mangal Complex, Near ICICI Bank, Kothrud, Pune – 411038, Maharashtra, India; Phone: +91 9890473801; Email: tejas@injotech.co.in

**Abstract:**

Liquefied Petroleum Gas is transported in bulk via ships, in liquid form. During transportation, various conditions cause the gas to vaporize. Reliquefaction systems are present to liquefy the same, consisting of compressors, condensers and associated piping. Cargo vaporizers are provided to vaporize the cargo as needed. These systems, consisting of shell and tube heat exchangers using sea water as either heating or cooling medium, are used to maintain pressures and to condition the tanks before operations. Tube failures in these exchangers can lead to sea water ingress into the cargo system, further leading to consequences including damage to machinery, complete shutdown and increase in downtime. NDE of these exchangers helps in monitoring their condition, prediction of future failures, preparing for docking, and planning for maintenance, while reducing downtime, risks of emergencies, and losses. Various NDE methods may be effectively used for inspection and examination of such heat exchangers.

**Keywords:** Eddy Current Testing (ECT), Remote Field Testing (RFT), Internal Rotary Inspection System (IRIS), Tube to Tube Sheet Radiography (TTS), Vacuum Box, Mass Spectrometer Leak Detection System (MSLDT), Liquefied Petroleum Gas (LPG), LPG Carriers.

## 1.0 Introduction:

Liquefied Petroleum Gas (LPG) is usually transported in bulk via ships dedicated for the same (LPG Carriers). LPG consists of a variety of gases including Propane, Butane, Ammonia, Mix, Propylene, Butadiene, Butene and Ethane. These gases are stored in dedicated cargo tanks on board ships, in liquid form, usually under high pressures and low temperatures. Due to the continuous movement of the liquid during transportation, the liquid tends to vapourize, thereby causing increase in tank pressures. These pressures need to be controlled for safety reasons and thus, there is present on board a reliquefaction system for the same. Also, in some cases, it may be required to vaporize the liquid on purpose so as to maintain adequate tank pressures to ensure the tank does not get under-pressurized. For this reason, cargo vaporizers are also provided on board. Both, the cargo reliquefaction and vaporizing systems, consist of heat exchangers of the shell and tube type, utilizing liquefied or vapour cargo on one side, and cooling or heating sea water on the other. Any failure of any of the tubes could lead to disastrous consequences, thereby, making regular inspections in-situ by Non-Destructive Examination of utmost importance.

## 2.0 Cargo System:

The cargo system consists of cargo tanks, reliquefaction plant, deep-well cargo pumps and associated valves and piping. The reliquefaction plant further consists of the cargo compressor, cargo condenser, intercooler and associated valves and piping. The cargo system also consists of a cargo vaporizer, used to vaporize liquid cargo for heating and pressure generation purposes. While the entire system is large and of great capacity, for the purpose of this paper, the system description and consideration has been restricted to the following items only.

1. Cargo Condenser
2. Cargo Vaporizer
3. Cargo Piping

Note: Refer to Fig.1 for a general arrangement of the Cargo System.

***2.1 Cargo Condenser:***

The cargo condenser is a shell-and-tube type heat exchanger. It is a part of the reliquefaction plant and is used to liquefy any LPG vapours that are generated during transportation. The general operation of the condenser system is as follows:

Vapours generated in the tanks are sucked in to the compressor, where they are compressed to a high pressure over multiple stages, with or without intercooling. After the final stage, the compressed gas is sent to the shell side of the condenser. The cooling medium is sea water at ambient temperature, pressurized by means of dedicated sea water pumps, which flows through the tube side. Liquid outlet from the condenser is sent back to the cargo tank via a condensate line.

The details of the condenser are as under:

Type: Shell-and-Tube Type

Shell Side Fluid: LPG (Vapour form)

Tube Side Fluid: Sea Water

Tube Arrangement: Straight Tube

Tube Material: Pure Titanium (98%), Titanium Alloy, Stainless Steel or similar

Tube Sheet Material: Low Carbon Titanium Alloy Steel

Tube Connection: Welded

Number of Tubes: Depends on size and capacity

Note: Refer to Fig. 2 for details of the Cargo Condenser.

**Figure 1: General Arrangement of LPG Cargo System**

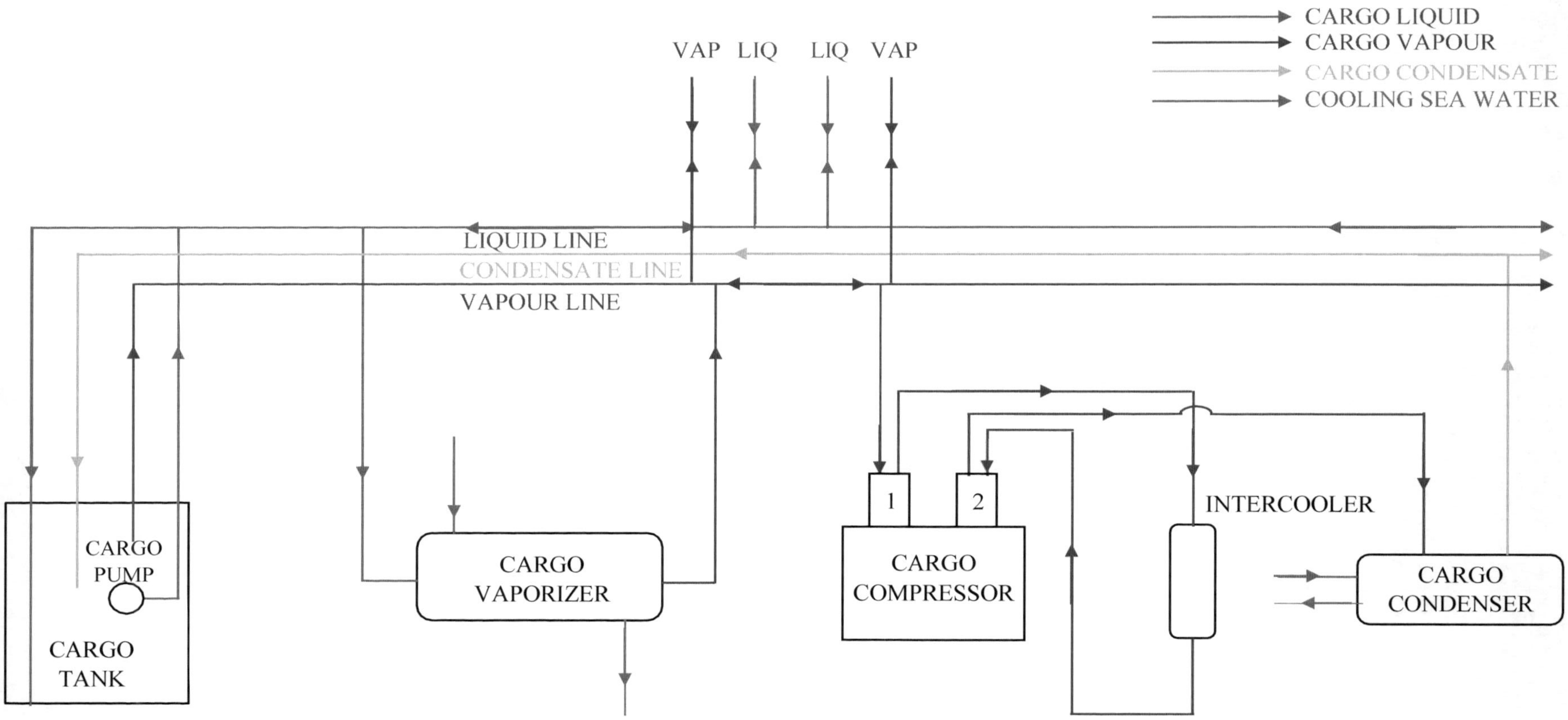

**Figure 2: Cargo Condenser**

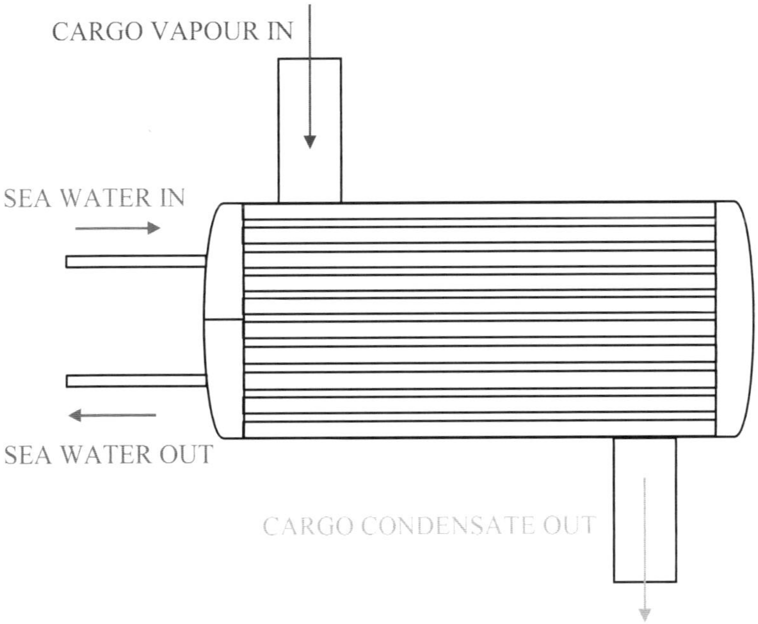


***2.2 Cargo Vaporizer:***

The cargo vaporizer is a shell-and-tube type heat exchanger. It is used to vaporize liquid cargo in order to control and maintain tank pressures. Tank pressure may need to be increased in cases of too low pressures, which may lead to implosion of the tank. Likewise, pressures may also need to be increased to assist in discharge of cargo, whereby vapour pressure acting on top of the liquid assists in pumping the liquid with suction from the tank bottom. The general operation of the vaporizer system is as follows:

Liquid cargo is supplied to the tube side of the heat exchanger. Sea water at ambient temperature is passed along the shell side. The sea water temperature, being higher than the normal cryogenic temperature of the cargo, heats up the cargo liquid, thereby converting it into vapours. These vapours are then led back to the concerned tank, where they are fed in from the top of the tank.

The details of the vaporizer are as under:

Type: Shell-and-Tube Type

Shell Side Fluid: Sea Water

Tube Side Fluid: LPG (Liquid form)

Tube Arrangement: U-Tube

Tube Material: Pure Titanium (98%), Titanium Alloy, Stainless Steel or similar (Plain Carbon steel may be used in certain heaters if temperatures are suitable for its use)

Tube Sheet Material: Low Carbon Titanium Alloy Steel

Tube Connection: Welded

Number of Tubes: Depends on size and capacity

**Figure 3: Cargo Vaporizer**

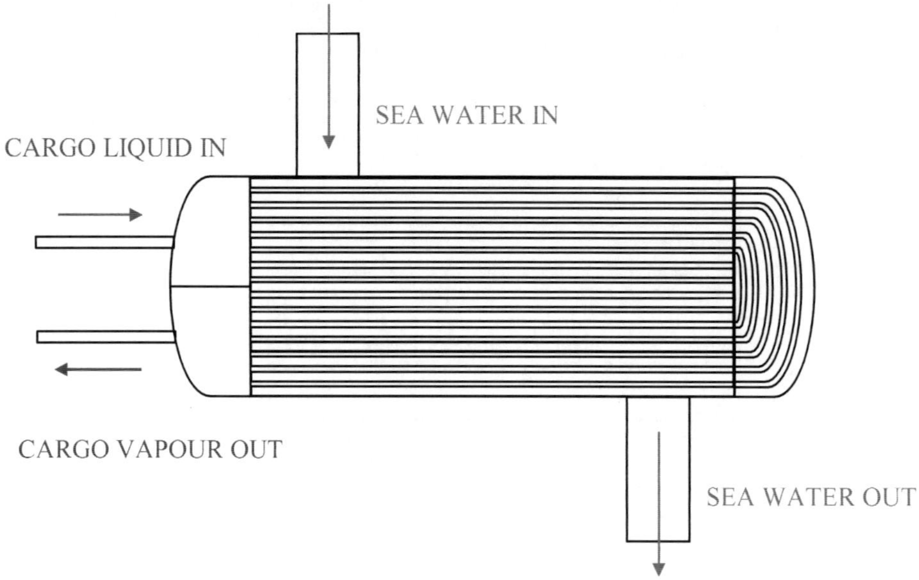


### *2.3 Cargo Piping:*

The cargo piping consists of various pipelines, valves, pumps and manifolds so as to transfer cargo to and from the various tanks, and load or discharge to the shore. The cargo manifolds are present on both, port and starboard sides, and usually consist of two (2) lines each for liquid and vapour. Condensate is never pumped ashore and thus, there is no condensate manifold. The liquid manifold is used to load or discharge liquid cargo, while the vapour manifold is used to receive vapour during discharging and to send vapour during loading in order to maintain tank pressures. Cargo pumps and the associated valves are used to pump cargo the manifold during discharging. Various safety valves are also provided to protect tanks from both over and under pressure. Refer to figure 1 for the general piping diagram of the cargo system.

## 3.0 Possible Problems:

The following problems can be expected in the condenser and vaporizers:

1. Pin-holes in the tubes
2. Cracked tubes
3. Corrosion of the tubes
4. Damaged welds between tube and tube sheet
5. Pitting
6. Support Wear

## 4.0 Consequences of Tube Failure:

Tube failures (or any of the problems mentioned above), would lead to sea water leaking into the cargo side. In the vaporizer, the cargo is at atmospheric pressure and the heating sea water is pressurized to a high pressure by means of pumps located in the engine room.

Similarly, in the condenser, while the cargo condensate is at a higher pressure due to being compressed in the compressor, this pressure is still quite low as compared to the sea water pressure. The sea water for both vaporizer and condenser is pumped via the same pumping and piping system, and thus, all parameters for the sea water are same for both. Also, usually, only one of the condenser or the vaporizer can be used at a time. Changing over from one to the other requires a complete stoppage and complete changeover of the sea water system.

Sea water leakage into the cargo side would have disastrous consequences. In case of the vaporizer, the water would be carried over into the vapour line and will be transferred to the tank. Likewise, in case of the condenser, the water would be carried over to the tank through the condensate line. The cargo in the tanks and in the condensate line being in a sub-zero cryogenic condition, would cause the water to freeze in the piping. Not only this, even a small amount of water being carried over (few drops) would freeze in any orifices or nozzles that are present in the line and inside the tank. (Spray nozzles are located inside the tanks for spraying liquid cargo before loading for tank preparation. The spraying vaporizes the cargo and brings down the tank temperatures to loading temperatures.)

The blockages in the piping and the nozzles would lead to severe pressures in the piping. This would cause excessive back pressures on the cargo compressors. The compressors are provided with excessive back pressure safety trips, which would then be activated, tripping the entire system. Tank preparation, loading, discharging and other cargo operations would be severely affected, leading to monetary losses, including ships going off-hire.

Due to the cryogenic conditions of the cargo, it is not possible to simply open up the piping and clear the water. Also, since small drops of water can accumulate in the nozzles and orifices, and these are quite often located inside the tanks, it is impossible to clear these. Normal tank entry would not be possible unless the cargo is completely emptied, and even then, entry would require inerting (filling the tanks with inert gas), gas-freeing and aeration operations, and proper enclosed space entry procedures. Likewise, reloading cannot occur until the reverse process of inerting and gassing up is carried out. As such, it would be a time consuming and monetarily expensive process, with long downtimes.

The only way to rectify a leak on board is to plug the concerned tubes. This not only requires a great deal of time and effort, leading to increased downtime, but is also temporary. Given the size of the leak, it may even be impossible to determine a leaky tube with the equipment and facilities available on board.

In short, the consequences can be summarized as follows:

1. Tube failures lead to water ingress in the tanks and piping.
2. Water freezes in the piping and nozzles due to cryogenic nature of cargo.
3. Freezing blocks nozzles and orifices, some of which are located inside the tanks.
4. This causes excessive back pressures on the compressors and cargo system.
5. Safeties get activated and the entire system trips.
6. Removal of water and restarting the system is a time-consuming process, and could lead to long downtimes and monetary losses.

## 5.0 Why NDE?

NDE is required to monitor the condition of the vaporizer and condenser tubes. While it may not be possible to carry out NDE at sea, any suspected leaks can be identified at the earliest and appropriate measures can be taken. Also, NDE would enable determination of remaining tube wall thickness and enable prediction of any future failures. Ships usually enter dry docks on a two-and-a-half yearly basis, thereby, NDE would enable the ship management to plan for the dry docking, and take corrective actions during the docking so as to eliminate the chances of tube failures in the future, by determining which tubes are faulty and which ones are close to being worn down or corroded.

In short, NDE can be carried out for the following reasons:

1. Determine exact tube and location for suspected leaks.
2. Determine tube thicknesses to predict future failures.
3. Eliminate chances for future tube failures by taking timely actions at dry docks.

## 6.0 NDE Methods:

The following NDE Methods can be applied for inspection of the condenser and vaporizer:

1. Eddy Current Testing (ECT)
2. Remote Field Testing (RFT)
3. Internal Rotary Inspection System (IRIS)
4. Tube-to-Tube Sheet Radiography (TTS)
5. Vacuum Box Bubble Leak Testing (BT)
6. Mass Spectrometer Leak Detection Testing (MSLDT) (Shell Side)

### *6.1 Eddy Current Testing (ECT):*

Eddy Current Testing using bobbin probes can be carried out on the tube side of the heat exchangers. Titanium or Titanium Alloys, being non-magnetic, are suitable to be tested with Eddy Current Testing. Similarly, Austenitic Stainless steel, which is predominantly the type of stainless steel used in low temperature applications, is also non-magnetic, thereby making the use of Eddy Current Testing suitable. Eddy Current Testing is not suitable for tubes made of Carbon or Alloy Steel, due to the fact that these types of steel are ferromagnetic in nature. ECT has a flaw detection capability of over 90% for non-magnetic materials, and around 66% for ferromagnetic materials. Overall, ECT has a good detection probability for almost all defects such as pitting, cracks, tube wear, support wear, and large and small volumetric flaws.

*Features:*

1. Suitable for examination of non-magnetic (diamagnetic) or slightly magnetic (paramagnetic) tubes.
2. Can detect and size corrosion, erosion, pitting, cracking, fretting, wear or a combination of the above.
3. Highly sensitive to ID/OD cracking, microbiological induced cracking (MIC) and small volumetric indications.
4. Poor sensitivity for ferromagnetic materials.

### *6.2 Remote Field Testing (RFT):*

Remote Field Testing (RFT) can be carried out on the tube side of the heat exchangers. Although the tubes are mostly made of Titanium and Titanium Alloys, or Austenitic Stainless Steel, in some cases, they may even be made of Carbon Steel or Low Alloy Steel. Such materials are rarely used, however. In such cases, RFT provides a suitable means of testing the ferromagnetic tubes. Due to the non-magnetic nature of Titanium or Stainless Steel, RFT is not suitable for use with such tubes. While RFT can be used for ferromagnetic materials, the flaw detection capability is only around 67% for ferromagnetic materials, and even lesser for non-magnetic materials.

*Features:*

1. Suitable for examination of ferromagnetic materials only.
2. Can detect ID and OD degradation such as corrosion, erosion, pitting, fretting and gradual wall loss.
3. Good sensitivity to vibration and steam erosion/impingement.
4. Poor sensitivity for non-ferromagnetic materials.

### *6.3 Internal Rotary Inspection System (IRIS):*

Internal Rotary Inspection System or IRIS is a type of ultrasonic examination that uses an arrangement of a water driven turbine and mirror to direct an ultrasonic beam into the material being tested. IRIS is suitable for use with ferromagnetic materials such as Carbon Steel. IRIS may also be suitable, to some degree, for examination of Titanium and Titanium Alloys, although the sensitivity is not as good as that for ferromagnetic materials. The sensitivity of IRIS for ferromagnetic materials is around 85%, while that for Titanium is around 70% and for Stainless Steel is less than 30%. Overall, IRIS has a good detection capability for support wear, moderate detection capability for pitting, tube wear and large and small volumetric flaws, and poor detection capability for cracks.

*Features:*

1. A type of Ultrasonic Immersion Technique.
2. Highly sensitive for ferromagnetic materials.
3. Can be used as a standalone examination method, or as a supplementary method for other tests.
4. Detects general thinning such as corrosion and erosion and localized defects such as pitting and wear scars.
5. Poor sensitivity for non-ferromagnetic materials.

### *6.4 Tube to Tube Sheet Radiography (TTS):*

Examination of the welding between the tube and the tube sheet, and to some extent the expanding, can be carried out by means of conventional film radiography. In this technique, a conventional film with a hole in the centre is used. The source is introduced inside the tube from the film side. The film is placed on the outside of the tube sheet and is covered by a lead shield. The back radiation technique is used, whereby, the back radiation from the source passes through the tube and tube sheet joint and then falls on the film. The film is developed in the conventional way to get an image of the joint in question. All other parameters, are controlled similar to conventional radiography.

**Figure 4: Tube to Tube Sheet Radiography Arrangement**

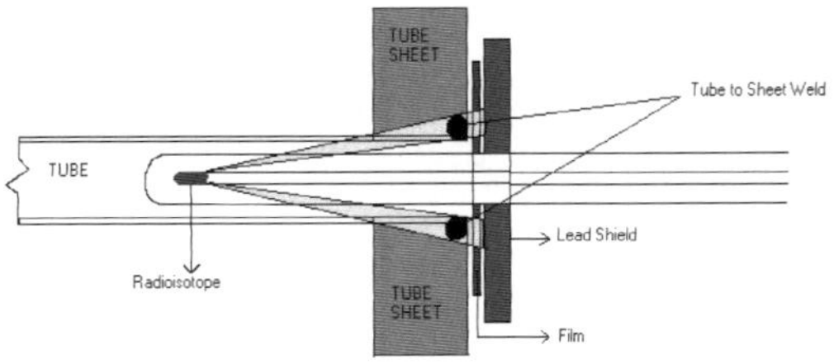

*Features:*

1. Technique similar to conventional radiography.
2. Uses back radiation technique.
3. High skill and experience required.
4. Danger of radiation exposure to health and safety.
5. Time consuming as each tube has to be separately radiographed.

***6.5 Vacuum Box Leak Testing:***

Vacuum Box Leak Testing is a type of differential pressure technique that can be used for examination of the welding between the tube and tube sheet. A vacuum box and a vacuum pump system is used to generate vacuum inside the box. A detergent solution is then applied. Any leaks are made visible via detergent bubbles created through the envelope. Leaks in welds due to through thickness discontinuities can be detected. The box is applied on the tube sheet, thereby enabling detection of any leaks in the welds and expanding of the tubes.

**Figure 5: Typical Vacuum Box Arrangement**

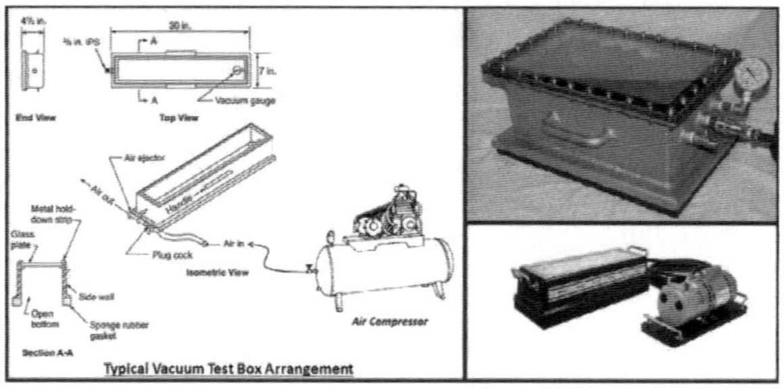


*Features:*

1. Provides an immediate visual indication of the location of any leak.
2. Small leaks can be detected.
3. Time consuming for large size of heat exchangers.
4. Secondary analysis may be required by other NDT methods.
5. Sensitive to testing temperature.

***6.6 Mass Spectrometer Leak Detection Testing (MSLDT):***

Mass Spectrometer Leak Detection Testing (MSLDT) can be used for testing the shell side of the heat exchanger. Not only can it be used to detect any leaks on the shell side, it can also be used for pressure testing of the shell side, if the conditions so allow. The shell side is isolated from the entire system. It can then be pressurized by helium gas, and any leaks from the shell can be measured. Shell side integrity is critical, especially in the cargo condenser as it carries LPG cargo on the shell side. Any leaks would cause pressurized LPG to either leak into the compressor room, or air to go inside the system. LPG being a mixture of toxic and combustible gases, a leak would be hazardous for the safety of any personnel inside the room. Conversely, the compressors are designed for handling LPG, and any air leakage would severely affect the working and may even damage the compressors.

*Features:*

1. Variable sensitivity available.

2. Difficult to develop pressure.
3. Good sensitivity to small leaks.
4. Sensitivity and accuracy may vary depending on conditions.

## 7.0 Conclusion:

NDE of Cargo Condensers and Vaporizers on LPG ships is critical. Tube failures in such systems lead to drastic and disastrous consequences. The onset of any defects or corrosion can be easily detected well in advance. This would enable the concerned management to plan for future dockings and repairs, and would also prevent problems for shipboard staff in their operations. It would also help reduce downtime. NDE of such condensers can be carried out in a variety of ways, however, the more convenient methods include Eddy Current Testing (ECT) for diamagnetic or paramagnetic materials, Internal Rotary Inspection System (IRIS) for ferromagnetic materials, Vacuum Box Bubble Leak Testing (BT) for testing the integrity of tube to tube sheet joints and Mass Spectrometer Leak Detection Helium Leak Testing (MSLDT) for shell side inspection.

## 8.0 Acknowledgements:

The author would like to thank his esteemed colleague Mr. Ganesh Jagtap, from Injotech India Pvt. Ltd., for his help with the NDT related knowledge and information in regards to this paper. The author would also like to thank his previous superiors on board various Gas Carriers, during his sailing career, for their guidance and assistance during his time with them, for enabling him to understand the working of the cargo systems. While it would be difficult to acknowledge everyone by name, the contribution made by each one, in any form, knowingly or unknowingly, has gone in some way in helping to make this paper.

## 9.0 References:

1. Hyundai Heavy Industries Co., Ltd., Ulsan Shipyard, Korea – Process Flow Diagram, Cargo System for LPG Carrier.

2. Burckhardt Compression AG, Winterthur, Switzerland – Operation Manual for LPG Labyrinth Piston Compressor.

3. DongHwa Entec, Korea – Operation Manual for Cargo Heater/Vaporizer.

4. Birring, A., NDE Associates Inc. – Eddy Current Testing vs. Ultrasonic IRIS for Inspection of Heat Exchanger Tubing.

5. Dey, Anup Kumar, whatispiping.com – Vacuum Box Testing of butt, fillet and lap weld.

The e-Journal of Nondestructive Testing - ISSN 1435-4934 - www.ndt.net

# Basic Study on Impact Elastic Wave Method to Expand Application Scope of Grouting Condition Evaluation in PC Tendon Ducts

Kentaro YAMASHITA, Shinichi HATTORI, Koki TERASAWA, and Toshiro KAMADA

Division of Global Architecture, Graduate School of Engineering, Osaka University , 2-1 Yamadaoka, Suita, Osaka 565-0871, Japan

Phone: +81-6-6879-7618, Fax +81-6-6879-7618; e-mail: k.yamashita@civil.eng.osaka-u.ac.jp, hattori_shinichi@civil.eng.osaka-u.ac.jp,terasawa@civil.eng.osaka-u.ac.jp, kamada@civil.eng.osaka-u.ac.jp

**Abstract**

This paper shows the possibility to extend the scope of application for evaluate grouting condition in PC tendon ducts, of conventional two types of elastic wave input methods.

At first, we focused on the difference between waveforms measured at symmetrically arranged points across the input point. By using elastic wave propagation simulation and experiments using full-scale specimens, we found that the location of the internal void could be detected from the changes in the amplitude or phase of the difference in waveforms at the two measurement points.

The second, we considered new indexes and effective sensor location, about the method which using electromagnetic input method to generate elastic wave. We found the phase characteristics, amplitude changes, and Akaike's information criterion (AIC), can be effective indexes to evaluate the grouting conditions in PC tendon ducts.

**Keywords:** PC grout, non-destructive evaluation, impact-echo method, electromagnetic impact method

## 1. Introduction

The Impact-echo method [1] involves generating and transmitting elastic waves into concrete by tapping concrete surface with a steel ball. Then evaluating the response signal of the measured elastic waves in the frequency domain, assuming that the measurement object has a structure that provides a single resonant frequency, i.e., a plate shape. When elastic waves are input to a concrete element whose cross-section is not plate-shaped (e.g., a bottom flange of a prestressed concrete (PC) bridge girder), multiple unexpected frequencies are generated as the elastic waves propagate along a complex path. Furthermore, even if the measurement object has a plate-like cross-sectional shape, stable multiple reflections of elastic waves may not be established between a circular-shaped (unfilled) tendon duct and the measurement surface; hence, in some cases, it is difficult to evaluate the grout filling condition in a part that does not have a plate-shaped cross-section, using a method based on frequency analysis.

Since most of the concrete structures used in social infrastructure are not plate-shaped, the scope of application of the impact echo method is limited. Therefore, technical development to expand its scope of application is necessary.

In this study, we investigated two methods of elastic wave input to expand the scope of application of grout condition evaluation by elastic waves, namely, the "mechanical" method by tapping a steel ball, and the "magnetic" method by using pulsed magnetic field to excite vibration in tendon duct and PC steel.

When the grout filling rate of tendon ducts in a structure is small, it is the same as when voids are present in the concrete. They become free-surface reflectors against the propagating elastic waves, resulting in the generation of reflected waves. Therefore, if the reflected wave from the void can be detected, it will be possible to estimate the location and size (filling rate) of the void by focusing on the arrival time and the strength of the reflected wave. However, the direct detection of the reflected wave from the voids has not been established in the use of the impact echo method so far. One of the reasons for this is the effect of the surface waves generated when tapping the concrete surface. The surface waves have large amplitudes and separation of the reflected waves from internal voids becomes difficult when their amplitude is small. Therefore, the authors are investigating a technique to measure vibrations at positions symmetrical and equidistant from the signal input point, and to detect reflected waves from internal voids by removing the effect of the surface waves using the difference of the waveforms. [2]

On the other hand, in the magnetic input method, an exciting coil is installed on the concrete surface directly above the buried part of the tendon duct, and a magnetic field is generated around the coil by applying a pulsed current, which vibrates the tendon duct and PC steel in the concrete, thereby inputting elastic waves into the concrete. The main feature of this method is that we can make the evaluation object, i.e., a duct or PC steel, the source of elastic waves. The magnetic input method has the potential to evaluate the grout filling condition from a different perspective than the conventional method because it is not affected by the noise generated when the concrete surface is tapped and because the elastic waves generated by directly vibrating the tendon can be measured. In previous studies, the effectiveness of an evaluation method using the amplitude information of elastic waves measured at the concrete surface by vibrating a tendon with magnetic input has been reported[3], and this method has been applied to actual structures. However, since the amplitude of elastic waves is affected by attenuation in concrete, and it is only possible if the concrete cover thickness of the tendon duct and the propagation path are constant, we have not yet reached the stage where we can evaluate the amplitude of elastic waves with high accuracy regardless of these conditions.

## 2.Method focused on waveform difference in mechanical inputs

### *2.1 Principle of measurement*

If there are no reflective surfaces such as the edge of the material or internal defects in the vicinity of the two measurement points placed symmetrically at the same distance from the input point of the elastic wave, then the waveform of the same phase and amplitude will be measured, and the difference will be zero. By contrast, as shown in Fig. 1, if the other signal than input such as a reflected wave from a void are overlapped at a different timing at the output point, this can be detected as the difference between the two waveforms. In addition, when detecting the reflected wave from the internal void at the concrete surface, it was essential to eliminate or reduce the influence of the surface wave that has a larger amplitude than the reflected wave. Assuming that the surface wave propagates in concentric circles from the input point, this can be canceled by the difference of the measured waveforms at two measurement points located at the same distance.

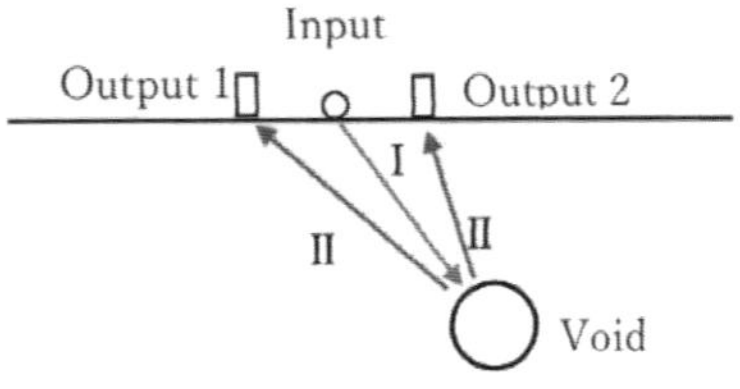


Fig. 1 Schematic illustration of measurement the waveform difference

However, when the signal input point is directly above the void, the reflected wave from the void is also erased by the difference. Therefore, it is necessary to perform a sweep measurement by moving the measurement unit consisting of the signal input point and two measurement points, but the changes in phase and amplitude of the waveform difference can be used as an indicator to detect the unfilled grout.

### *2.2 Analytical study*

The evaluation method of unfilled grouts based on waveform difference was investigated by numerical calculation using the finite element method. The model of calculation simulates an unfilled tendon duct as shown in Fig. 2, with a circular void 50 mm in diameter and a concrete cover thickness of 50 mm or 100 mm. The response waveforms were calculated by moving the measurement unit, with the distance between the input point and the two output points (L) set at 50 mm, between 200 mm on either side of the void at intervals of 25 mm. The input was a 10 kHz pulse signal, and the calculation time interval was 0.5 ▯s for 1000 steps. As a result of the calculation, it was confirmed that the amplitude difference between the waveforms of the two output points became larger under the condition that the measurement point was located near the void. Fig. 3 shows the relationship between the position of the input point andthe difference in the maximum amplitude of the wave front. The amplitude difference is zero when the input point is directly above the void, and the amplitude difference reverses to an extreme value when one of the measurement points is located near the directly above the void (±50 mm position). It was confirmed that the arrival of reflected waves from voids, i.e., unfilled ducts, could be detected without using frequency analysis.

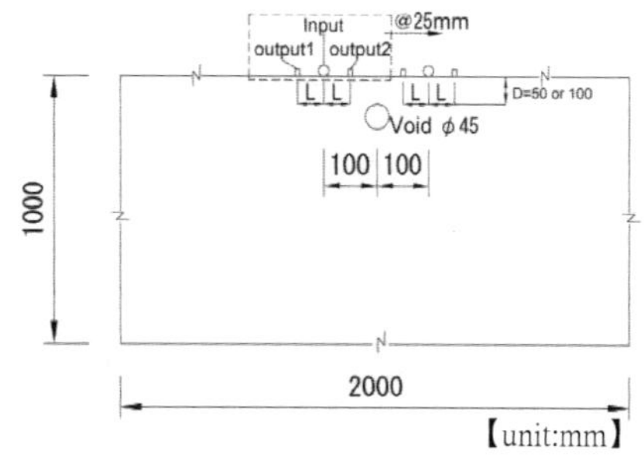

Fig. 2 Model of numerical calculation

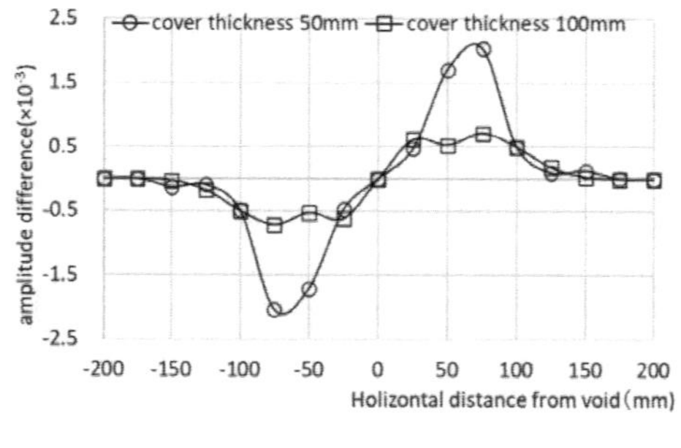

Fig. 3 Amplitude change of waveform difference

### *2.4 Full-scale specimen experiment*

### *2.4.1 Outline of specimen*

The appearance of the full-scale concrete specimen of the PC girder is shown in Fig. 4, where the grout filling condition in the ducts at the bottom flange is changed at each 1.5-m section within the total length of 7.5 m. In this study, four combinations of filling rates for the overlapping ducts of No. 4 and No. 5, shown in Fig. 5, are investigated. The compressive strength of the concrete is 40 N/mm$^2$, and the tendon duct is made of steel with an inner diameter of 45 mm. The 100% filled ducts were grouted in a vertical position beforehand, and placed in the formwork after curing.

Fig. 4 Full-scale PC girder specimen

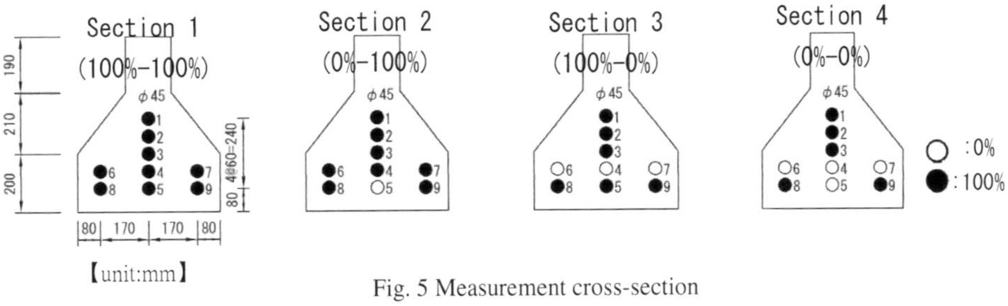


Fig. 5 Measurement cross-section

## *2.4.2 Measurement and analysis methods*

In the impact elastic wave method, elastic waves are generally input by lightly tapping the surface of concrete with a steel ball or the like. Since the input wave is affected by the concrete properties, and there is a possibility of the input position may be slightly deviated, the frequency properties of input wave is not constant. Therefore, it would be a problem for detailed evaluation on the existence and location of internal voids. As a solution to this problem, we have experimentally confirmed that by tapping a steel ball set on the input point with a steel ball of similar mass provides a more stable signal input than direct tapping at the concrete surface. As shown in the measurement overview in Fig. 6, input was carried out by tapping a □15-mm steel ball in contact with the input point with an impulse hammer of similar mass, and the input waveform were recorded. The measurement unit, consisting of the input point and two measurement points (accelerometers), was moved at intervals of 25 mm over an area of 75 mm in front and behind the No. 5 duct. The waveform difference was standardized by dividing the difference between the maximum amplitudes of the initial motion of the velocity waveforms measured at the two measurement points by the amplitude of the input waveform. Waveforms were recorded for 5.0ms at a sampling interval of 1.0□s.

Fig. 6 Measurement overview (left: whole image, right:close-up)

### *2.4.3 Measurement results of full-scale specimen*

Fig. 7 shows the changes in the waveform difference obtained from the full-scale specimen. The horizontal axis of the figure shows the plane distance from the duct of No. 5, and the vertical axis shows the standardized amplitude difference value. The figure shows that the difference in amplitude is almost zero in the 100%–100% combination, while in the condition where there is an unfilled tendon duct in one of the two stages, a point-symmetrical pattern of extremes is observed at ±50 mm, where one sensor is directly above the duct. The results show that by focusing on the amplitude difference of the waveform measured by the mechanical input method, it becomes possible to evaluate the grout filling at the bottom flange, where application of conventional methods was difficult. On the other hand, it was found to be difficult to determine the filling ratio of the back side duct in the case where the filling ratio of the front side duct was 0%. We consider this was due to the fact that if there is an unfilled duct on the front side, then the elastic waves are reflected there and cannot reach the duct on the back side.

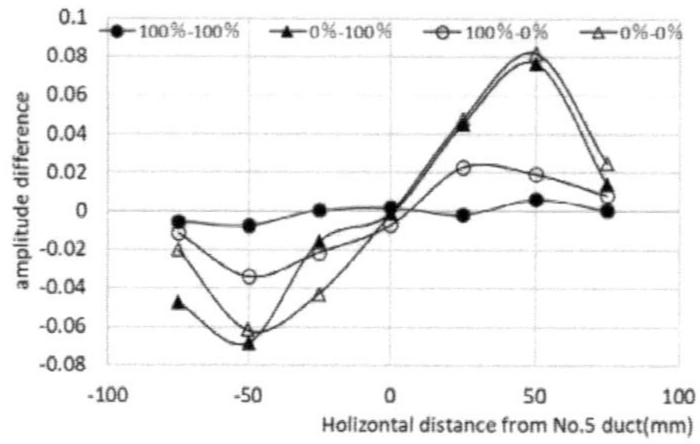


Fig. 7 Waveform difference calculation results on full-scale specimen

## 3.Magnetic input method

### *3.1 Principle of measurement*

Fig. 8 shows a schematic of the magnetic input method. When an electric current is applied to the exciting coil installed on the concrete surface directly above the tendon duct using a pulse power supply, a magnetic field is generated from the exciting coil, and an impact (electromagnetic force) can be applied to the steel duct and PC tendon in the concrete, which are magnetic materials. The elastic waves generated by this impact propagate through the concrete and are received by sensors (such as AE sensors) installed on the concrete surface.

Since the generated magnetic field acts without being affected by voids, previous investigations have confirmed that weak electromagnetic force can be applied to the ducts on the back side even when the ducts are placed in overlapped positions in the depth direction. Therefore, there is a possibility to evaluate the grout filling condition of the tendon ducts placed in two stages, which was difficult by the mechanical input.

Fig. 9 shows the outline of electromagnetic input device. Electrical steel sheet with high permeability and large magnetic flux density at saturation are laminated in a U-shape to generate an large magnetic flux. The coil cross-section is 50mm x 50mm, and the magnetic path length is 300mm. The number of excitation coil turns are eight turns per end.

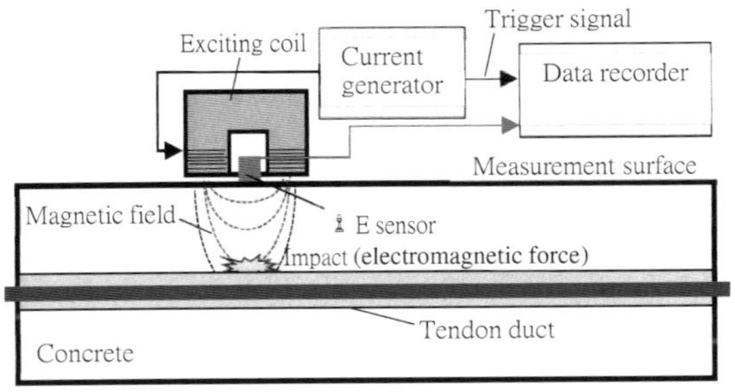


Fig. 8 Schematic of the magnetic input method

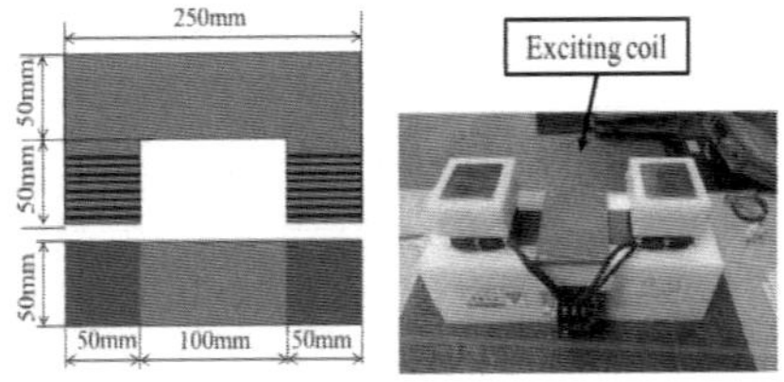


Fig. 9 Outline of exciting coil

If the mechanical impedance of the contact area between the tendon duct and the grout is $Z$ and the electromagnetic force acting on it is $F$, then the vibration velocity $V$ of the generated vibration is expressed as follows:

$$V = \frac{F}{Z} \quad \text{.......................................................} \quad (1)$$

With the magnetic input method, the force (electromagnetic force) is constant, so the magnitude of the vibration generated is governed by the state of the mechanical impedance. If the solidified grout behaves as an elastic body, and the degree of spring coefficient $k$ per unit contact area is constant, then the realized spring coefficient $K$ for contact area $A$ is given by

$$K = kA \quad \text{.......................................................} \quad (2)$$

Therefore, if the larger the contact area is, then the larger the spring coefficient becomes.
Also, since the relationship between the spring coefficient $K$ and the mechanical impedance $Z$ is expressed by the following equation.

$$Z = \sqrt{mK} \text{.......................................................} (3)$$

Where $m$ is the mass per unit length of the duct.
Furthermore, as the frequency of the excited vibration $\omega$ is given by

$$\omega = \sqrt{\frac{K}{m}} \quad \text{.......................................................} \quad (4)$$

If the larger the contact area is, the larger the grout filling ratio is, then the frequency of the vibration generated by the tendon duct will be higher and the amplitude will be smaller; therefore, grout filling condition may be grasped by focusing on the frequency and amplitude values of the measured vibration.
It should be noted that in the previous study, it was reported that the sum of squared amplitudes tends to decrease as the grout filling ratio increases, which is consistent with this hypothesis.

*3.2 Consideration of evaluation index values by numerical calculation*

### *3.2.1 Model of calculation*

Impact response analysis was performed using the finite element method with the input value of electromagnetic force acting on the duct obtained by three-dimensional dynamic magnetic field analysis. The analysis was conducted in four patterns of conditions, which is multiplication of two ducts installed in the model with a thickness of 300 mm, width (duct length) of 500 mm, and length of 2000 mm as shown in Fig. 10, and two grout fill rates of 0% and 100%.
The input point of the impact force (electromagnetic force) was divided into nine sections at intervals of 50 mm in the direction of the extension of the duct, and each section was provided with four nodes (top, bottom, left, and right), thus 36 points. The calculation intervals were set to 1.0 ▯s and performed 5000 steps. For simplicity, the PC tendon in the duct were not set in the model.

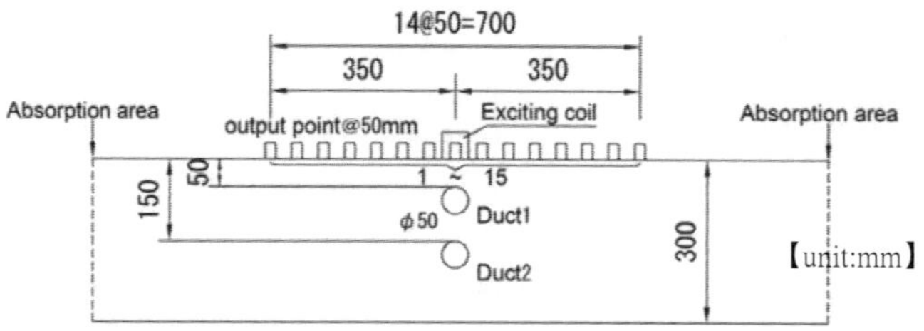


Fig. 10 Finite element analysis model

### *3.2.2 Waveform amplitude index*

Fig.11(a) shows the relationship between sum of the squares of the amplitude values of the obtained calculated waveforms and the filling rate. It was confirmed that the sum of squared amplitudes was larger when the filling ratio of duct 1 was 0%, but the effect of the filling ratio of duct 2 was hardly reflected in the sum of squared amplitudes index. This was considered to be due to the larger electromagnetic force acting on the duct closer to the measurement surface. In addition, there was no significant change in the waveform by the output position.

### *3.2.3 Information amount of waveform*

In general, for a system that oscillates only at a particular frequency, the motion is predictable, and thus the volume of information of the signal in that state is small and the information entropy is high. On the other hand, if there are multiple vibration systems and it is not possible to determine which is dominant at a given time, the amount of information is large and the information entropy is low. When an unfilled tendon duct is vibrated by electromagnetic force, irregular vibrations may occur in several vibration systems compared with the fully grouted condition. Since this vibration is considered to be less predictable and with a larger amount of information, we tried to investigate the possibility of evaluation of filling using the change of information amount as an index. In this study, we used the Akaike Information Criterion (AIC), which is also used in seismic wave analysis, as an index of information amount. Fig. 11(b) shows the results of finding the minimum value of AIC for the waveform calculated at each output point. From the figure, it was confirmed that the amount of information of the excited vibration changed in accordance with the grout filling ratio of duct 1, which was closer to the measurement surface, and it showed the applicability of the evaluation index, but the effect of the filling ratio of duct 2 was hardly reflected. It was also found that the difference in the amount of information was larger for the measurement points on the side than those directly above the duct.

### *3.2.4 Instantaneous frequency*

As mentioned earlier, the frequency of vibration generated in ducts may vary depending on the grout filling ratio; therefore, a study using the frequency of vibration generated in ducts as an index was conducted. For frequency analysis method, Fourier Transform or Fast Fourier Transform (FFT) are commonly used; however, these methods assume that the target signal is stable (in a stationary stochastic process with a long duration), and it is difficult to capture frequency changes that occur in a small amount of time. Also, there is a concern that the information desired to be obtained may be buried due to the influence of the shape of the concrete part or the multiple reflections of elastic waves by using a waveform taken over a long period of time.

Here, since the amount of phase change of a signal of any frequency is constant per unit time, it is possible to determine the frequency (instantaneous frequency) at any given time from the amount of phase change. The instantaneous frequency is an index that is also used in the analysis of signals with large non-stationarity and shock loads. In this study, the phase change of the measured waveform was obtained by the Hilbert transform and calculated from the amount of change per unit time.

$$f = \frac{1}{\Delta T}\frac{\Delta\phi}{2\pi} \quad \text{............................................................} \quad (5)$$

where $f$ : Instantaneous frequency, $\Delta\phi$ : Amount of phase change, $\Delta T$ : Any given time

Fig. 12 shows the maximum value of the instantaneous frequency of the waveform at each measurement point. From the figure, changes of the instantaneous frequency according to the filling ratio was observed at the measurement points located on the side rather than those near the duct, which indicated the effectiveness of the two-stage duct as an indicator of the filling ratio.

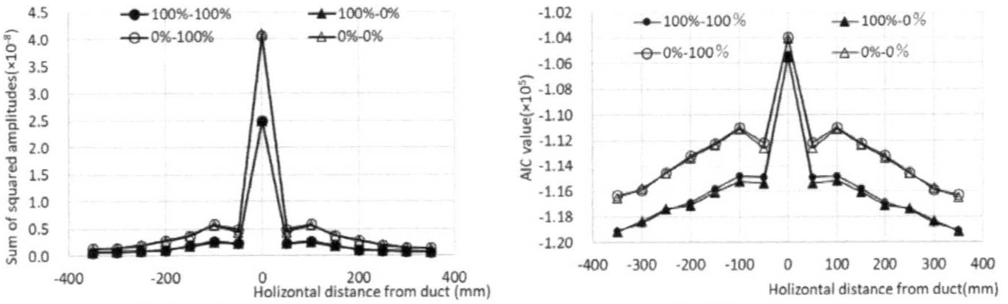


(a) Sum of squared amplitudes (b) AIC minimum value

Fig. 11 Index calculation results

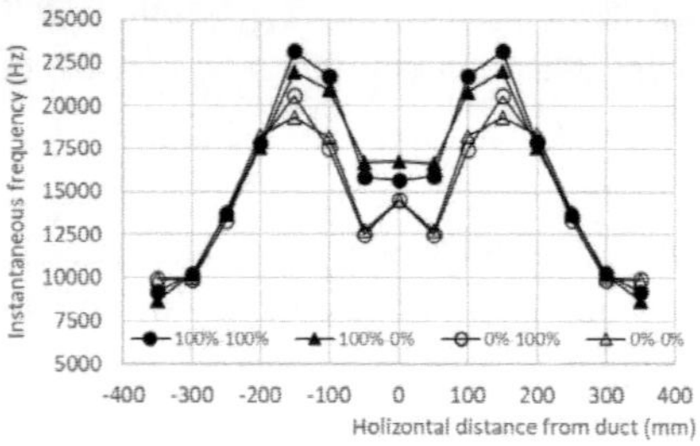


Fig. 12 Instantaneous frequency calculation results

### *3.3 Full-scale PC girder specimen experiment*

#### *3.3.1 Measurement conditions*

The measurements were performed on ducts No. 4 and No. 5 of the four cross-sections of the full-scale specimen described in 2.4.1. As shown in the Fig. 13 measurement overview, the exciting coil was placed at a distance of 10 mm from the concrete surface directly above the No. 5 duct, and the elastic wave response was measured by four AE sensors attached with double-sided tape at intervals of 50 mm from the exciting coil. Waveforms were recorded for 5.0ms at a sampling interval of 1.0?s.

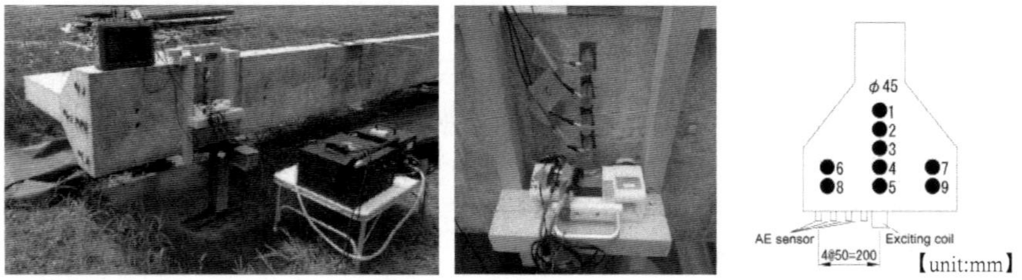


Fig. 13 Measurement overview and sensor position

#### *3.3.2 Measurement results*

The index values at each measurement point obtained by the magnetic input method are shown in Fig. 14 and Fig. 15. For both indexes, the same trend as in the numerical calculation was confirmed at measurement point 1, which was 50 mm from the input point. For measurement points 2 to 4, the changes in the index values are small, which is considered to be due to the effects of attenuation of elastic waves, etc.

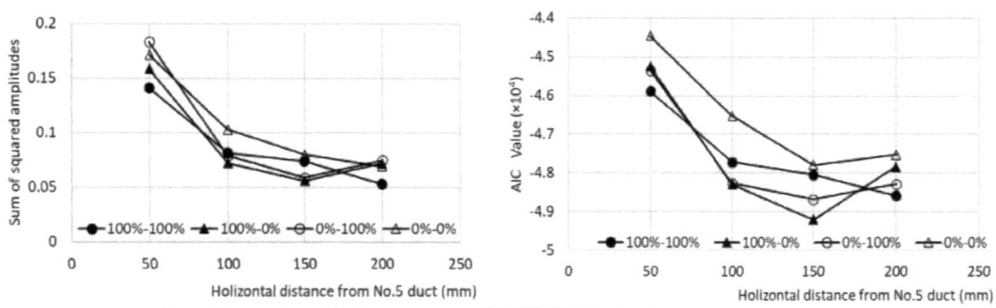


(a) Sum of squared amplitudes (b) AIC minimum value

Fig. 14 Measurement results

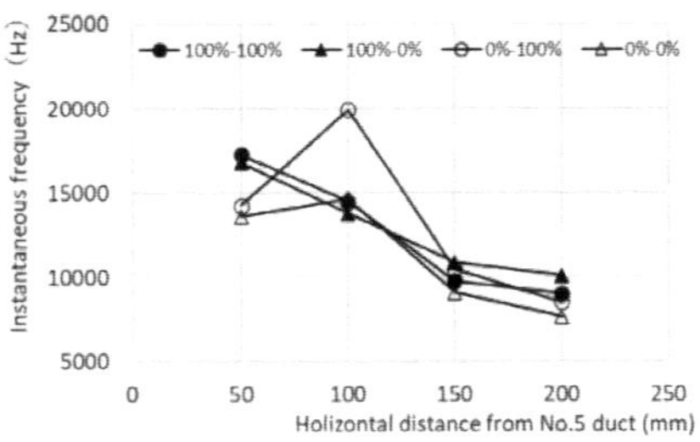


Fig. 15 Measurement results: Instantaneous frequency

## 4. Conclusions

- The applicability of the method for detecting unfilled grouts was confirmed by measuring the changes in amplitude difference of waveforms received by sensors located at positions symmetrical and equidistant from the input point, and mechanically inputting pulsed elastic waves. However, it was difficult to evaluate the filling condition on the back side where multiple ducts overlap.
- In addition to the sum of squared amplitude that has been used as an index for the magnetic input method, another applicable evaluation index was found.
- The analytical study of the magnetic input method showed that there were conditions for positioning of the measurement points depending on the index, but the full-scale specimen experiment showed that the measurement points closest to the duct best reflected the filling condition. For practical use, it is also necessary to consider the attenuation of elastic waves.
- As a result of the numerical calculation and the measurement of the two-stage section by the magnetic input method, changes of several index values according to the combination of the filling ratio were observed. This indicated the possibility of evaluating the filling condition by statistical methods such as multivariate analysis.
- Compared with the magnetic input method, the mechanical input method requires a simpler measurement system, with no power supply and less time for measurement, but it is difficult to obtain information on the back side of a two-stage duct. Therefore, a combination of the two methods, e.g., screening by the mechanical input method and confirming any anomalies found by the magnetic input method, will further expand the applicability to PC grout filling evaluation and improve the evaluation accuracy.

### References

1. Mary J. Sansalone William B. Streett; *IMPACT-ECHO Nondestructive Evaluation of Concrete and Masonry*, 1997

2. Yamashita, K. Tezuka, M. Kamada, T '*Basic study on PC grout filling evaluation focusing on the difference waveform obtained by pulse input of elastic waves' Proceedings of the Japan Concrete Institute,* Vol.43, No.1, pp1241-1246, 2021

3. Hattori, S. Kamada, T. Uchida, S. Asakura, H. Terasawa, K. '*Basic study on the relation between elastic wave input methods and detection capability of imperfectly grouted part of tendon ducts in PC members', Proceedings of Japan Society of Civil Engineers,* E2, Vol. 73, No. 2, pp. 239-250, 2017

4. Hattori, S. Kibe, T. Terasawa, K. Kamada, T. '*A basic study of non-destructive evaluation method for detection of grouting condition in tendon ducts of PC members integrating mechanically and electromagnetically activated elastic wave together with electromagnetic response of steel members in concrete', Proceedings of the Concrete Structure Scenarios, The Society of Materials Science, Japan,* Vol.18, pp.243-248, 2021

# Automated defect detection in CFRP using flash infrared thermography and deep learning method

Zongfei TONG [1,2], Saeid HEDAYATRASA [1], Liangliang CHENG [1], Shejuan XIE [2], Mathias KERSEMANS [1]

[1] Mechanics of Materials and Structures (UGent-MMS), Department of Materials, Textiles and Chemical Engineering (MaTCh), Ghent University, Technologiepark-Zwijnaarde 46, 9052 Zwijnaarde, Belgium
[2] State Key Laboratory for Strength and Vibration of Mechanical Structures, Shaanxi Engineering Research Center of Nondestructive Testing and Structural Integrity Evaluation, Xi'an Jiaotong University, 710049 Xi'an, China

**Abstract**
Carbon fiber reinforced polymer (CFRP) composites have become an important material for many industry applications due to their high specific stiffness and specific strength. Inevitably, defects generated during manufacturing and/or in-service process could compromise the structural health of the CFRP components. Flash infrared thermography(IRT) is a promising NDT technique, which has already been successfully applied for the detection of various defects in a range of materials. In order to bring this technique to the next level, this study aims to accomplish automatic detection and localization of delaminations in CFRP using flash IRT experiments and deep learning-based object detection method. A virtual dataset has been generated by means of a custom developed fast numerical simulator (programmed in Fortran) of flash IRT. Then, a pre-trained object detection framework from literature, i.e. Faster-RCNN, is applied to the virtual dataset using the concept of transfer learning. A comparison with classical post-processing methodologies, e.g. thermographic signal reconstruction, and principal component thermography, is presented. Finally, a CFRP slab including twelve artificial rectangular delaminations was inspected using flash IRT and evaluated through trained Faster-RCNN.

**Keywords:** CFRP, Non-destructive testing, Flash infrared thermography, Numerical simulation, Object detection, Transfer learning

## 1. Introduction

Composite materials are manufactured by combining two (or more) materials, in view of exploiting the benefits of both, and as such to outperform classical materials. As one of the promising composite materials, carbon fiber reinforced polymer (CFRP) have become an important material for many industrial applications due to its high specific stiffness and strength. Unfortunately, defects generated during manufacturing and/or in-service process could compromise the structural health of the CFRP components. Therefore, non-destructive testing (NDT) techniques are needed to detect defects as early as possible, and as such to assure the structural safety and long-term reliability of CFRP components [1].

Although a variety of characterization and inspection techniques (optical, vibrational, ultrasound, eddy current, etcetera) have already been demonstrated, each of them has its own limited capability and applicability limitations [2]. Flash infrared thermography (IRT) is a promising NDT technique, which is capable of realizing non-contact and quasi real-time measurements over a large detection area [3]. In flash IRT, a short but intense optical flash excitation is utilized to introduce a heat flux on the surface of the inspected sample, which drives the diffusion of thermal waves in the material's through-thickness direction. Meanwhile, the surface temperature distribution and evolution are recorded with a sensitive infrared camera. The thermal diffusivity mismatch between any potential defect and the surrounding sound region results in both a locally disturbed spatial temperature distribution and an abnormal temporal temperature evolution [4]. As such, a defect can be revealed as a disturbance on spatial temperature distribution or as a localized anomaly in the temporal temperature evaluation profile. In order to quantitatively evaluate defects from a sequence of thermographic images, the key point is to extract proper temporal and spatial features.

In recent years, deep learning-based methods, gain a lot of attention in the fields of medicine, industry and architecture in order to extract higher level features from input images [5]. The frameworks based on convolutional neural network (CNN) show a promising potential for the accurate and automated defect detection in IRT [6]. However, the feature-rich thermographic image sequences make it challenging to extract temporal and spatial features simultaneously. Many scholars consider the thermographic sequences as the array of temperature series and use full connected NN [7], RNN[8], 1D CNN[9], AE[10] to extract the temporal features of temperature series. The novel framework based on Generative Adversarial Networks (GAN) has been proposed to extract the spatial features from several frames [11]. In order to extract temporal features and spatial features at the same time, three different strategies have been proposed. Ref. [12] assumes that the temporal and spatial features are not related, the temporal features and spatial features are extracted through full connected NN and object detection architectures, respectively. In contrast to ref. [12], the temporal features are compressed in ref. [13] by PCA to enhance the detection capacity of Faster-RCNN. Finally, a hybrid of spatial and temporal deep learning architecture is proposed, which adopts segmentation frameworks and long short term memory (LSTM) network to extract spatial features and temporal features, respectively [14]. From the references mentioned above, the temporal features of thermographic image sequences not only can be used to estimate defect depth, but also to evaluate defect shapes and size, while the spatial features can only be adopted to estimate defect shapes and size. One of the common challenges is that a completely sufficient experimental training dataset is difficult to achieve in practice. In this study, the numerical dataset was constructed using a fast numerical simulator. A traditional thermographic image processing method will be utilized as the pre-processing procedure for deep learning method. Among the traditional image processing methods, the principal component thermography (PCT), thermographic signal reconstruction (TSR), pulsed phase thermography (PPT) and the derivatives of these methods are effective strategies to highly compress the 3D data without losing too much information. By applying deep learning-based object detection methods, the automatic detection and localization of delaminations in CFRP can be accomplished.

The first challenge is to train the object detection framework by using a large and diverse set of IRT data. Setting up a complete experimental training database is difficult to achieve in practice. Therefore, a virtual database has been generated by means of a custom developed fast numerical simulator (programmed in Fortran) of flash IRT. The discontinued nodes method was utilized to simulate delamination-like defects in CFRP. Then, the PCT and TSR were exploited to construct the thermographic image dataset for deep learning-based object detection framework. Finally, a pre-trained object detection framework from literature, namely Faster-RCNN [15], was applied to the virtual database using the concept of transfer learning [12]. The structure of the paper is as follows: Section 2 introduces the developed fast numerical simulator. Section 3 presents the performance of Faster-RCNN on different simulated thermographic datasets and one experimental result. The conclusions are reported in section 4.

## 2. Fast numerical simulator for flash IRT applied on CFRP

### *2.1 Basic formulation*

The process involving the flash IRT applied to CFRP is a heating conduction issue [6]. Considering the energy conservation and Fourier heat conduction processes, the governing equation of the heat conduction can be expressed as,

$$\rho c\frac{\partial T}{\partial t}=\nabla\cdot\left(\bar{\bar{k}}\nabla T\right)+Q,\quad \text{in}\,\Omega \tag{1}$$

where $T$[K], $\varrho$[kg/m³], and $c$[J/(kg·K)] are the temperature, the density, and the specific heat, respectively. $\bar{\bar{k}}$ denotes the thermal conductivity matrix of anisotropic media. $Q$ denotes the heat generation function per unit volume. Ω represents the domain which needs to be solved.

The sample surface is subjected to thermal flux, which is stimulated by the excitation source of flash IRT. The thermal flux applied on sample surface can be expressed as,

$$-k\frac{\partial T}{\partial n}=q_{flux}\quad \text{on}\,\Gamma_1 \tag{2}$$

where the $q_{flux}$[ W/m²] denotes heat flux density in the outer normal direction on the boundary $\Gamma_1$.

In this case, the convection boundary condition, i.e., Robin boundary condition, is considered. The convection boundary condition can be presented as,

$$-k\frac{\partial T}{\partial n}=h(T-T_f)\quad \text{on}\,\Gamma_2 \tag{3}$$

where $h$[W/(m²·K)] indicates convection heat transfer coefficient, $T_f$ represents the temperature of the surrounding fluid, $T$ is the temperature of the solid which needs to be calculated, $\Gamma_2$ denotes the boundary at which the Robin condition applies.

In the finite element technique, the continuous variable $T$ is approximated by $\dot{T}$ in terms of its nodal values through 'shape functions'.

$$\dot{T}=[N]\{T\} \tag{4}$$

Using Galerkin's method, Eq. (1) can be discretized and expressed in matrix form as,

$$[K]\{T\}+[C]\left\{\frac{\partial T}{\partial t}\right\}=\{F\} \tag{5}$$

where the stiffness matrix $[K]$, damping matrix $[C]$, and load vector $\{F\}$ can be presented as,

$$[K]=\sum\left(\int_{\Omega}\left(\nabla\{N_i\}\right)^T\bar{\bar{k}}\left(\nabla\{N_i\}\right)\mathrm{d}v+h\int_{\Gamma_3}\{N_i\}^T\{N_i\}\mathrm{d}s\right) \tag{6}$$

$$[C]=\sum\rho c\int_{\Omega}\{N_i\}^T\{N_i\}\mathrm{d}v \tag{7}$$

$$\{F\}=\sum\left(\sum h\int_{\Gamma_2}\{N_i\}^T T_f\mathrm{d}s-\sum\int_{\Gamma_1}\{N_i\}^T q_{flux}\mathrm{d}s\right) \tag{8}$$

Equation (5) can be solved using Crank-Nicholson strategy, finally leads to,

$$\left([K]+\frac{2[C]}{\Delta t}\right)\{T(t)\}=\{F(t)+F(t-\Delta t)\}+\left(\frac{2[C]}{\Delta t}-[K]\right)\{T(t-\Delta t)\} \tag{9}$$

The solution of Eq. (14) is the degree of freedom $T(t)$ at a certain time t.

In order to efficiently simulate delamination in CFRP, the discontinued nodes (DN) method was introduced in this study, which is represented in Fig. 1. The thickness of delamination is assumed to be zero in DN method. In comparison to simulate delamination using volume element, the DN method reduces the amount of element numbers considerably, and as such accelerates the process of constructing a virtual database.

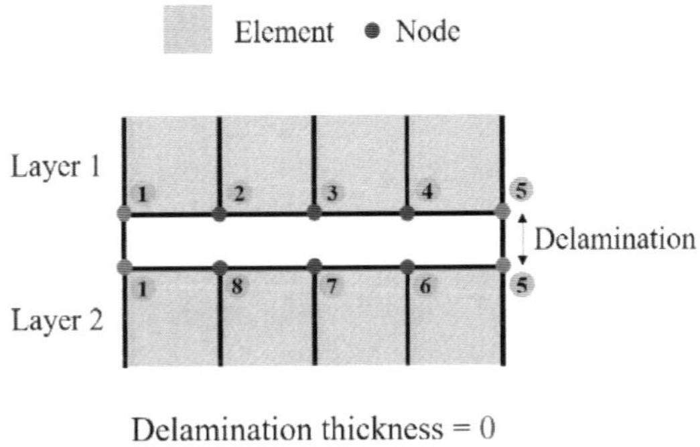


Figure 1 The diagram of discontinued nodes method

*2.2 Numerical model for constructing virtual database*

In order to establish the virtual database, a square CFRP plate model of size 150 × 150 × 1.84 $mm^3$ and quasi-isotropic material layup $[(0/+45/90/-45)]_S$ was adopted in this study as shown in Fig. 2. There is a delamination of variable size located in the middle of the part and its depth position iteratively changes. The size of the delamination and the material properties [17] are listed in table 1. A non-uniform Gaussian-shaped heat flux with 6 kJ is applied on the top surface of the sample, whose centre is randomly located on the sample surface. The heating stage lasts 10 ms, followed by a cooling stage of 14.99 s. The convection boundary condition has a convection coefficient equal to 5 $W/(m^2 \cdot K)$.

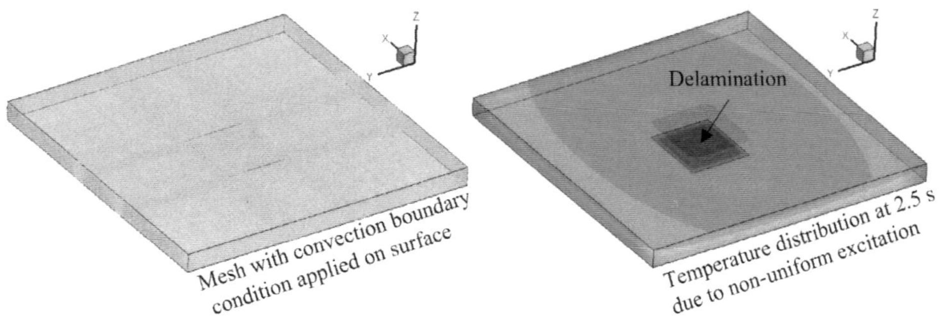

Figure 2 Numerical model

Table 1 Parameters of numerical model

| Quantity | Value |
|---|---|
| Density $\rho$ [kg/m³] | 1530 |
| Specific heat $c$ [kJ/(kg·K)] | 917 |
| Thermal conductivity $k$ [W/m·K] | Kx = 2.71, Ky = 0.61, Kz = 0.53 |
| Side length of the delamination with rectangle shape [mm] | 1.0 ~ 50.0 |
| Depth of the delamination [mm] | 0.23 ~ 1.61 |
| The number of numerical cases | 2975 |
| Simulation time | ≈ 4 min × 2975 cases ≈ 198.3 hours |

## 3. The performance of Faster-RCNN on synthetic datasets

*3.1 Thermographic image datasets for Faster-RCNN*

The entire time history of raw thermographic images obtained from simulation cannot be efficiently embedded into Faster-RCNN framework directly, and any individual time frame cannot represent all the features of 3D thermographic image sequences. Consequently, the PCA and TSR pre-processing methods are utilized to compress the temporal features of the thermographic image sequences. The temperature evolution of each pixel is standardized before conducting PCA and TSR, in order to remediate the effects of non-uniform heating [4]. The first four components of PCA are retained, considering that these amount for more than 95% of the variance in the original dataset. Similarly, the first four components of TSR are also adopted, while an eighth-degree polynomial is used to fit the temperature evolution profile. In order to consider the negative factors introduced by experimental errors to the datasets, the original image sequences are degraded with white Gaussian noise with a standard deviation 15 mK. Finally, four virtual databases are constructed: (1) PCA database, (2) PCA database with 15 mK noise, (3) TSR database and (4) TSR database with 15 mK noise. Each database consists of 11,900 images, of which 90% are used for training the Faster-RCNN, and 10% are used for testing. As an example , the processing results of a delamination with 10.0 mm × 10.0 mm and 1.61 mm depth are presented in Fig. 3.

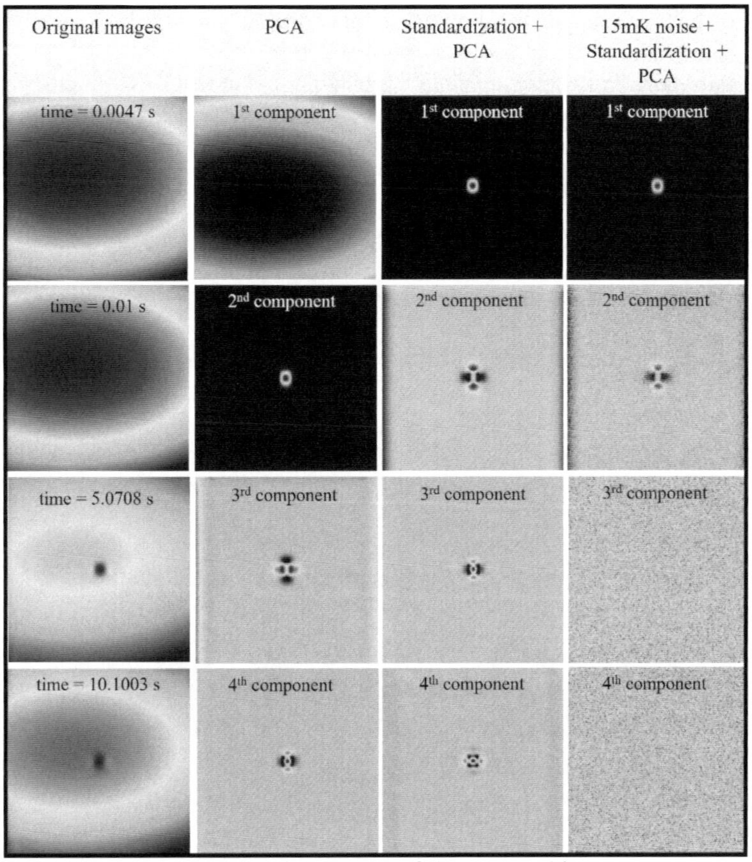


Figure 3 The processing results of thermographic sequences

*3.2 The performance of Faster-RCNN*

Faster-RCNN is, one of the most popular deep learning-based object detection algorithms [15], and has been employed to extract spatial features from thermographic image dataset in this study. Compared with other object detection frameworks, the superiority of Faster-RCNN is that it introduces a Region Proposal Network (RPN), which embeds region proposal directly in the architecture, and alleviates the need for a selective search algorithm. A pre-trained Faster-RCNN trained on COCO dataset [18] was fine-tuned through transfer learning. The base network of the Faster-RCNN is Resnet-101, which is used for classification task [19]. The anchor's scale ranges from 0.25 to 2.0, while aspect ratios of each anchor range from 0.1 to 10.0. The stochastic gradient descent (SGD) with a momentum term set to 0.9 was used in Faster-RCNN, whose initial learning rate is set at 0.0003. The training steps are 50000 and the batch size is set to 1. The training process and validation process (every 500 steps) cost about 6 hours on a RTX 2060 graphics card. The intersection over union (IoU) and mean average precision (mAP) [15] are adopted to evaluate the performance of Faster-RCNN in this study. The evaluation results are listed in table 2. The mAP of Faster-RCNN in validation dataset have promising

accuracy. The noise added to dataset has limited effect on the TSR results but decreases the inference accuracy of PCA results significantly.
Finally, a CFRP slab including twelve artificial rectangular delaminations was inspected using flash IRT [16]. The first TSR coefficient image was adopted to test the Faster-RCNN model trained on 'TSR + 15 mK noise' dataset (see table 2). From Fig. 4, it is clear that six of the defects are successfully detected. However, it can be seen that the results are heavily disturbed by the measurement noise and non-uniform background. Moreover, the defected areas are not necessarily rectangular and their thermal signature is significantly distorted. Therefore, the diversity of the training dataset (in terms of e.g., shape and orientation of defects) should be further improved for reliable detection of the irregular and hardly detectable defects.

Table 2 Performances of Faster-RCNN for different dataset after 50000 training steps

| Dataset | Loss | mAP@.50IoU | mAP@.75IoU | Average mAP |
|---|---|---|---|---|
| PCA | 0.00344 | 98.90% | 93.89% | 91.33% |
| PCA + 15 mK noise | 0.00491 | 93.30% | 85.60% | **84.32%** |
| TSR | 0.04085 | 97.97% | 93.87% | 91.12% |
| TSR + 15 mK noise | 0.01196 | 97.97% | 95.01% | **92.54%** |

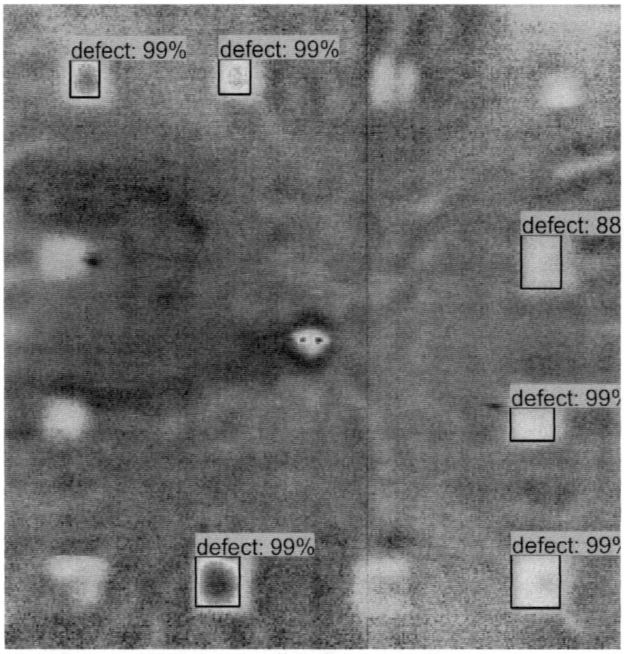


Figure 4 The inference result on the first TSR coefficient of experimental result

## 4. Conclusions

Deep learning-based object detection algorithm (Faster-RCNN) has been applied on pre-processed simulated thermographic data, in view of automated detection and sizing of delamination defects in CFRP laminates. The numerical dataset has been obtained using a custom-built fast simulator programmed in Fortran. The PCA and TSR processing methods were used to compress the temporal features of the 3D thermographic sequences, after which, the Faster-RCNN was employed to detect and size delamination defects. Although the noise decreases the inference accuracy of Faster-RCNN trained on PCA dataset, the PCA and TSR pre-processing methods are effective methods to compress temporal features of 3D thermographic sequences for deep learning-based architectures. Experimental examination of the trained algorithm reveals the significant impact of measurement noise and non-uniform background on detection quality and shows that the defects with irregular thermal signature cannot be detected. Hence, in the future, the numerical dataset will be enriched with a wide range of defect shapes and the methodology will be demonstrated on a sparse experimental dataset by considering a second transfer learning step on this object detection framework.

**Acknowledgements**
The authors acknowledge the financial support from the China Scholarship Council (CSC, grant number 202006280481).

**References**

1. R.A. Smith, Composite defects and their detection, Materials Science & Engineering, (1997).
2. M. Kersemans, I. De Baere, J. Degrieck, K. Van Den Abeele, L. Pyl, F. Zastavnik, H. Sol, W. Van Paepegem, Nondestructive damage assessment in fiber reinforced composites with the pulsed ultrasonic polar scan, Polymer Testing, 34 (2014) 85-96.
3. R. Yang, Y. He, Optically and Non-optically Excited Thermography for Composites: A Review, Infrared Physics & Technology, 75 (2016) 26-50.
4. G. Poelman, S. Hedayatrasa, J. Segers, W. Van Paepegem, M. Kersemans, An Experimental Study on the Defect Detectability of Time-and Frequency-Domain Analyses for Flash Thermography, Applied Sciences, 10 (2020) 8051.
5. I. Goodfellow, Y. Bengio, A. Courville, Deep Learning, MIT Press, 2016.
6. Z. Tong, S. Xie, X. Li, C. Pei, Z. Chen, Y. He, Efficient numerical simulation of eddy current pulsed thermography NDT signals based on FEM-BEM method and energy equivalent principle, Infrared Physics & Technology, 101 (2019) 138-145.
7. A. Darabi, X. Maldague, Neural network based defect detection and depth estimation in TNDE, Ndt & E International, 35 (2002) 165-175.
8. C. Hu, Y. Duan, S. Liu, Y. Yan, N. Tao, A. Osman, C. Ibarra-Castanedo, S. Sfarra, D. Chen, C. Zhang, LSTM-RNN-based defect classification in honeycomb structures using infrared thermography, Infrared Physics & Technology, 102 (2019) 103032.
9. Y. Cao, Y. Dong, Y. Cao, J. Yang, M.Y. Yang, Two-stream convolutional neural network for non-destructive subsurface defect detection via similarity comparison of lock-in thermography signals, NDT & E International, 112 (2020) 102246.

10. J. Xie, C. Xu, G. Chen, W. Huang, Improving visibility of rear surface cracks during inductive thermography of metal plates using Autoencoder, Infrared Physics & Technology, 91 (2018) 233-242.
11. L.F. Ruan, B. Gao, S.C. Wu, W.L. Woo, DeftectNet: Joint loss structured deep adversarial network for thermography defect detecting system, Neurocomputing, 417 (2020) 441-457.
12. N. Saeed, N. King, Z. Said, M.A. Omar, Automatic defects detection in CFRP thermograms, using convolutional neural networks and transfer learning, Infrared Physics & Technology, 102 (2019).
13. J. Hu, W. Xu, B. Gao, G.Y. Tian, Y. Wang, Y. Wu, Y. Yin, J. Chen, Pattern deep region learning for crack detection in thermography diagnosis system, Metals, 8 (2018) 612.
14. Q. Luo, B. Gao, W.L. Woo, Y. Yang, Temporal and spatial deep learning network for infrared thermal defect detection, Ndt & E International, 108 (2019).
15. S. Ren, K. He, R. Girshick, J. Sun, Faster R-CNN: Towards Real-Time Object Detection with Region Proposal Networks, Ieee Transactions on Pattern Analysis and Machine Intelligence, 39 (2017) 1137-1149.
16. B. Azizinasab, R.P. Hasanzadeh, S. Hedayatrasa, M. Kersemans, Defect Detection and Depth Estimation in CFRP through Phase of Transient Response of Flash Thermography, IEEE Transactions on Industrial Informatics, (2021).
17. C. Maierhofer, M. Röllig, M. Gower, M. Lodeiro, G. Baker, C. Monte, A. Adibekyan, B. Gutschwager, L. Knazowicka, A. Blahut, Evaluation of different techniques of active thermography for quantification of artificial defects in fiber-reinforced composites using thermal and phase contrast data analysis, International Journal of Thermophysics, 39 (2018) 1-37.
18. https://github.com/tensorflow/models/blob/master/research/object_detection/g3doc/tf1_detection_zoo.md
19. K. He, X. Zhang, S. Ren, J. Sun, Deep residual learning for image recognition, in: Proceedings of the IEEE conference on computer vision and pattern recognition, 2016, pp. 770-778.

# Active thermography to detect defects in wood samples of interest for cultural heritage

Massimo RIPPA[1], Pasquale MORMILE [1]

[1] Institute of Applied Sciences and Intelligent Systems "E. Caianiello" of CNR, Pozzuoli (Na), Italy

Phone: +39 0818675334, e-mail: m.rippa@isasi.cnr.it, p.mormile@isasi.cnr.it

**Abstract**
Non-destructive analysis is an important topic for the conservation of works of art. The development of tools based on advanced technologies and the study of their effective contribution to the analysis of materials is fundamental to guarantee the conservation and to monitor the state of health of the artistic heritage. Thermography is a well-known non-invasive technique for providing effective analysis of different types of materials. In the history of art, wood represents one of the most used materials to make different types of works such as panel paintings, sculptures, carvings and others. In this work we use Pulsed Thermography (PT) to detect artificial defects (air holes and metal inclusions) in samples of different wood essences. The results found are compared and both the advantages and limitations of such an analysis are discussed.
**Keywords:** Thermography, Non-Destructive Testing (NDT), wood, cultural heritage, pulse thermography.

## 1. Introduction

The conservation of works of art belonging to cultural heritage is a topic of great interest. In the case of artworks made using wood, such as panel paintings, statues, sculptures and carvings, various factors can affect their integrity: fluctuations in environmental parameters, the presence of pollutants, exposure to inadequate light or the formation of bacteria. These factors can change the colours or produce mechanical deformations which in turn cause cracks in the work and more generally increase its degradation. Detecting the formation of defective areas is the main goal of non-destructive inspection. Among the various investigation methods used for this purpose, Active Thermography (AT) is a well-known non-invasive imaging technique for providing a rapid and low-cost analysis of a work of art [1-4]. According to this technique, the surface of the object under investigation is stimulated using an external heat source to produce a dynamic thermal response that can be detected and recorded with an infrared camera. AT allows a rapid identification of a wide variety of defects such as cracks, voids, inclusions, delaminations, detachments, the presence of nails and any type of anomaly that determines a change in the thermophysical properties of the object under examination. This study reports the analysis related to the detection of defects in specimens of wood essences of interest for works of art through the use of the Pulse thermography (PT). PT is an active method in which the samples under investigation are thermally stimulated by means of the use of a heat pulse. Specimens of three different essences of wood: walnut, chestnut and fir, where defects such as air holes and metal inclusions of different sizes and depths were artificially made, have been investigated. The comparison of the results found shows how the detection limit can be very different and, in particular, strictly dependent on the specific essence of wood.

## 2. Materials and methods

*2.1 Wood specimens*

For the analysis, samples of three different essences of wood respectively of walnut, chestnut and fir (dimensions 8 cm × 8 cm × 2 cm) were considered. In each sample, an equal matrix of air holes (scheme showed in figure 1a), with diameters of 15, 12, 9, 6 and 3 mm and depth of 1, 2, 3 and 4 mm from the investigated front surface, was artificially made using a drill press. The depths of the holes were accurately obtained and checked using a micrometric translator coupled to the drill for moving the specimens in the z axis. Subsequently, the air holes made in the samples were filled with screws of the same size to simulate metal inclusions. In figure 1c, a picture of the rear surface of one of the wood specimens analysed is showed.

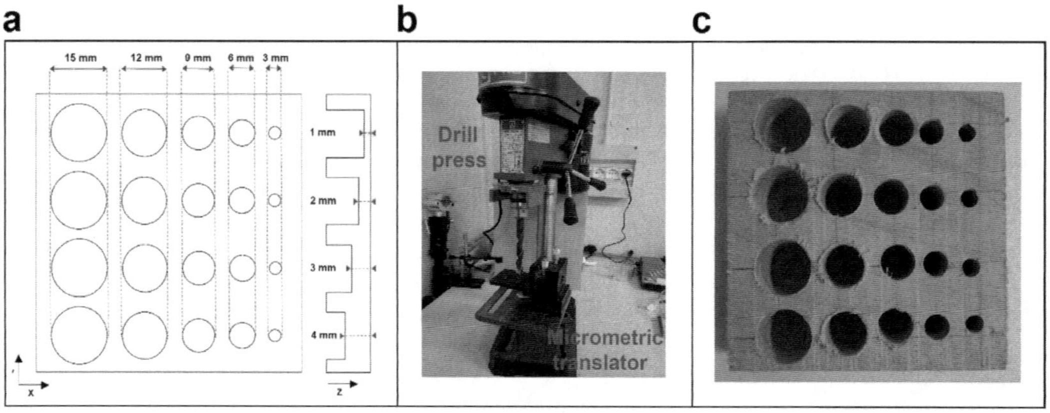


Figure 1. Realization of wood specimens: a) scheme of the matrix of the air holes considered, b) drill press coupled with a micrometric translator used to make air holes with different both diameters and depths in the wood samples, c) image of the rear surface of one of the wood specimens analysed.

*2.2 Pulsed Thermography measurements*

PT measurements were carried out in order to detect the presence of the defects within wood specimens. Two halogen lamps with tunable power were used as heat source to apply a thermal pulse of 40 s on the front surface of the samples investigated. The thermal response achieved during and after heating was recorded with a frame rate of 5 Hz using the infrared camera FLIR X6580 sc with a cooled indium antimonide (InSb) detector (FPA 640 × 512 pixels and NETD ~ 20 mK at 25 °C) mounting a 50 mm focal lens with spectral band 3.5–5 μm and IFOV 0.3 mrad. The software ResearchIR (FLIR Systems), was used for monitoring the temperature in real-time and for basic operations. The emissivity considered for all wood specimens was 0.90 [5]. The measurements were done in laboratory with 24 °C of ambient temperature and 50 % of relative humidity.

## 3. Results and discussion

The wood specimens made were investigated using the PT in order to evaluate its limit of detection for defects as air holes and metallic inclusions in the case of the three different essences taken into account. We note here that in the literature there are already several works where AT techniques are used to identify the presence of defects in the wood [6-16]. However, in these works, differently from our intents, no comparison is made between different essences and the results obtained refer to wood in general. From the infrared movie, recorded during and after the application of the heat pulse, were extrapolated the images where the thermal contrast between defectives and a non-defective reference area is highest and hence where the air holes under the front surface of the specimens are most clearly visible. In figure 2 are shown the images extrapolated for each wood essence analyzed.

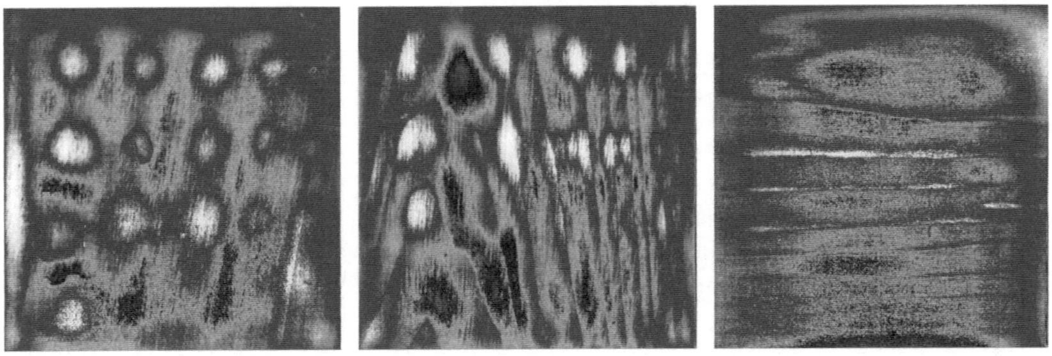


Figure 2. Thermal images with the highest defect visibility achieved using the PT analysis for the three wood essences investigated: a) walnut wood, b) chestnut wood and c) fir wood.

As visible from the images, in the case of the walnut specimen (left image in figure 2) the experimental method used allows to detect clearly all the air holes up to a depth of 3 mm. In addition, even the first hole with a diameter of 15 mm at depth of 4 mm is visible. In the case of chestnut (central image in figure 2), compared to the previous case, the detection limit is reduced and only the first two rows of air holes (up to 2 mm deep) are detectable. In the last case considered relative to the fir specimen (right image in figure 2), with the experimental approach used, none of the air holes present under the surface are visible. The results achieved for the three wood specimens analyzed highlight how even if both the structural texture and the thermo-physical properties characterizing the essences considered aren't so different, they can play a crucial role regarding the detection limits of the investigation method used in this work and, in general, in the success of infrared inspection.
Subsequently, the presence of metal defects in the specimens was simulated by filling the holes with screws of the same size. In figure 3, as an example, are shown the thermal images with highest thermal contrast achived with the PT analysis relative to the walnut specimen and in the case of both air and metallic defects.

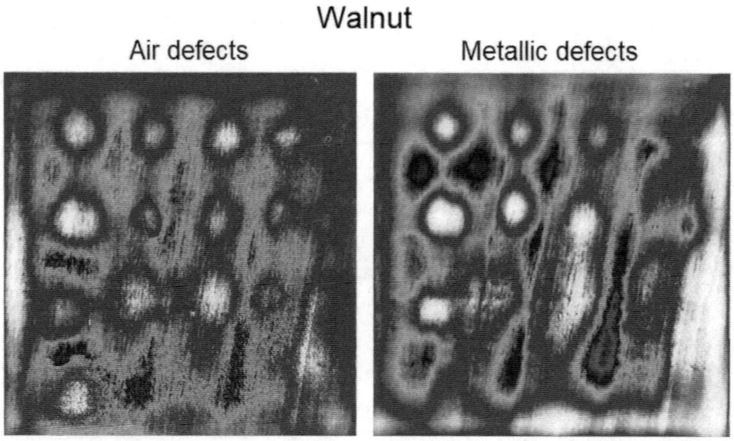


Figure 3. Thermal images with the highest defect visibility achieved using the PT analysis for the walnut specimen: a) air holes defects, b) metallic defects.

As can be seen from the images, the presence of metal inclusions reduces the visibility of the defects allowing to detect only some holes in the first two rows and only the large 15 mm diameter hole in the third row (depth 3 mm). This comparison demonstrates that the higher thermal conductivity of the metal screws doesn't favor the detection for the experimental approach used probably due to the faster way in which the heat is dissipated and expelled from the rear surface of the specimen.

## 4. Conclusions

The PT analysis showed in this work highlight how the type of wood essence plays an important role in obtaining an effective thermographic investigation and being able to identify the presence of the various defects within this material. Our results therefore demonstrate how a preliminary investigation on the type of wood to be investigated is essential, especially in the case of works of art, to predict the chances of success, take into account the limits of the analysis and better optimize the inspection parameters.

## References


1. X.P.V Maldague, 'Theory and practice of infrared technology for nondestructive testing', John Wiley & Son, New York, 2001.
2. V. Vavilov and D. Burleigh, 'Infrared Thermography and Thermal Nondestructive Testing', Springer Nature Switzerland AG, 2020.
3. M. Rippa, V. Pagliarulo, A. Lanzillo, M. Grilli, G. Fatigati, P. Rossi, P. Cennamo, G. Trojsi, P. Ferraro and P. Mormile, 'Active Thermography for Non-invasive Inspection of an Artwork on Poplar Panel: Novel Approach Using Principal Component Thermography and Absolute Thermal Contrast', J Nondestruct Eval, Vol. 40, No. 21, 2021.
4. P. Theodorakeas, S. Sfarra, C. Ibarra-Castanedo, N.P. Avdelidis, M. Koui, X.P.V. Maldague, D. Ambrosini and D. Paoletti, 'The use of Pulsed Thermography for the investigation of art and cultural heritage objects' 5th International Conference on NDT of HSNT- IC MINDT 2013, Athens, Greece.
5. G. López, L.A. Basterra and L. Acuña, 'Determination of the emissivity of wood for inspection by infrared thermography', J. Nondestr. Eval., Vol. 32, pp. 172–176, 2013.
6. S. Sfarra, C. Ibarra Castanedo, D. Ambrosini, D. Paoletti, A. Bendada and X. Maldague, 'Integrated Approach between Pulsed Thermography, Near Infrared Reflectography and Sandwich Holography for Wooden Panel Paintings Advanced Monitoring', Russ J Nondestruct Test, Vol. 47, pp. 284–293 2011.
7. A. Bendada, S. Sfarra, D. Ambrosini, et al., 'Active Thermography Data Processing for the NDT and E of Frescoes', QIRT 2010, July 27–30, 2010.
8. G. López, L.A. Basterra and L. Acuña, 'Infrared thermography for wood density estimation', Infrared Physics & Technology, Vol. 89, pp. 242-246, 2018.
9. R. Pitaforma, J. Crisostomo and L. Pereira, 'Detection of wood damages using infrared thermography', Procedia Computer Science Vol. 155, pp. 480–486, 2019.
10. D. Vidal and R. Pitarma, 'Infrared Thermography Applied to Tree Health Assessment: A Review' Agriculture, Vol. 9, No. 156, 2019.
11. Wyckhuyse and X. Maldague, 'Wood inspection by infrared thermography' IVth. IWASPNDE, X. Maldague ed., TONES (ASNT pub.), Vol. 6, pp. 201-206, 2002.
12. C.F. Bremer, G.N.A. Mello, A. Maracahipe, P.G. Krüger, 'Active thermography as non-destructive technique in the diagnosis of pathological manifestations in wood pieces', International Journal of Development Research, Vol. 8, No. 11, pp. 23883-23889, 2018.
13. M. Kucharska, J. Jaskowska-Lemanska, 'Active Thermography in Diagnostics of Timber Elements Covered with Polychrome', Materials, Vol. 14, No. 5, pp. 1134, 2021.
14. P. Meinlschmidt, 'Thermographic detection of defects in wood and wood-based materials', 14th international Symposium of nondestructive testing of wood, Hannover (Germany), May 2nd-4th 2005.
15. A. Wyckhuyse and X. Maldague, 'A study of wood inspection by infrared thermography, Part I: Wood Pole Inspection by Infrared Thermography', Research in NonDestructive Evaluation, Vol. 13, No. 1, pp. 1-12, 2001.
16. A. Wyckhuyse and X. Maldague, 'A study of wood inspection by infrared thermography, Part II: Thermography for Wood Defects Detection', Research in NonDestructive Evaluation, Vol. 13, No. 1, pp. 13-21, 2001.

# Digitalization for Railway-NDT

Daniel WERNER[1], Andreas FRANZEN[1], Frank HENRIX[1], Uwe PHILIPPS[1], Paul BUSCHKE[1]

[1] Waygate Technologies, Huerth, Germany, Phone: +49 2233 601115, Email: d.werner@bakerhughes.com, andreas.franzen@bakerhughes.com, frank.henrix@bakerhughes.com, uwe.philipps@bakerhughes.com, paul.buschke@bakerhughes.com

**Abstract**

High-speed trains, as they are used by leading train operators in Germany e.g., are inspected with underfloor testing machines. During inspection testing machine detects condition of wheels, including rim, tread, inner and external face and wheel disk, without disassembling wheel.

Due to its innovative testing procedure, high automatization level and user-friendly controlling and inspection tools, this third generation of inspection machine is a good example of how to use Industry 4.0 ideas in non-destructive testing.

To increase producibility inspection system allows inspection of a wheel in approx. 90 seconds. Inspection system supports operator with a straightforward workflow-based inspection procedure, several visualization and inspection tools. Internal, or external experts such as NDT competence center can support in case of difficult results without being on-site. Solutions for connectivity and tools for stand-alone-evaluation also increase flexibility of evaluation process. Thus, human factor can be reduced, and inspection reproducibility is maximized.

**Keywords:** Wheel inspection; in-service inspection; UFPE-WheelStar; rail; high speed trains; testing machine; high automatization level; ultrasonic testing (UT); NDT4.0; industry 4.0

## 1. Introduction

For highspeed trains several definitions exist, but they all have a minimum travelling speed of 190 km/h in common. Of course, many highspeed trains travel much faster than 190 km/h. Due to fast travelling speed safety is indispensable to avoid catastrophes.

Thus, leading train operators in Germany e.g., inspect wheels of highspeed trains regularly by using ultrasonic testing (UT).

Ultrasonic testing includes inspection not only of running surface, but of the complete wheel, including wheel disk, inner and external face, rim and tread (figure 1).

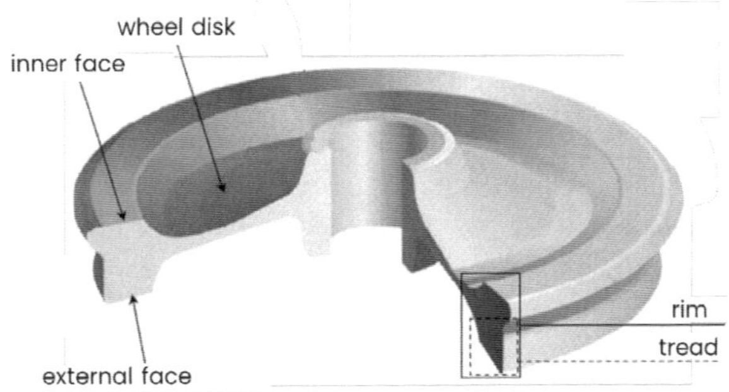


Figure 1. Areas inspected during in-service-wheel inspection

Areas shown in figure 1 are inspected in parallel to other maintenance work. Details on inspection areas are described in DIN 27201-7 [1].

Disassembling of the wheel is not necessary, so that inspection does not mess with maintenance work, which is relevant to realize inspection without blocking capacities of trains.

## 2. Underfloor inspection system

Therefor underfloor testing machines drive in a pit under the train. Figure 2 gives a schematical view to describe inspection process:

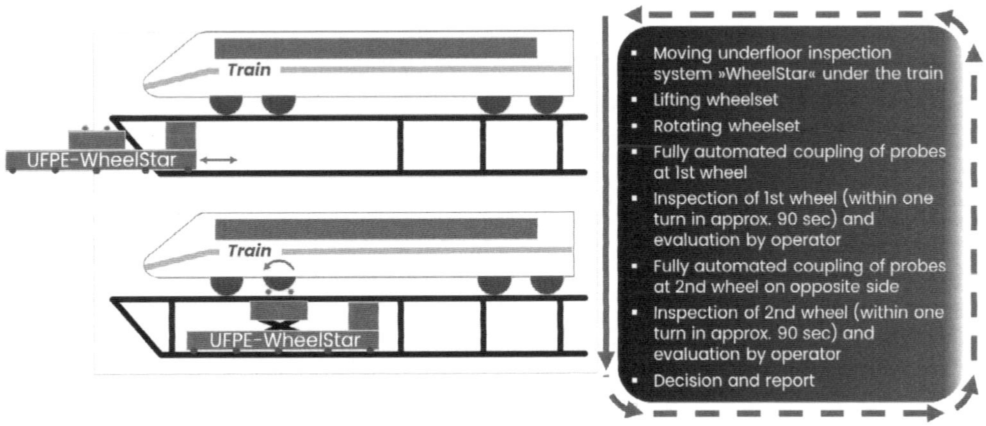


Figure 2. Underfloor testing machines for-wheel inspection (in principle) and testing procedure

First, underfloor inspection system is moved to the wheelset to be inspected. This wheelset is lifted and rotated by a hydraulic system integrated in the underfloor inspection system. Next, probes are coupled to the wheel. After a wetting turning (water is used for coupling) inspection of the first wheel starts. Lifting and rotating of wheel as wheel as coupling of probes is fully automated.

Within one turn (in approx. 90 sec) wheel is inspected (incl. rim, tread, inner and external face and wheel disk). Operator evaluates testing results by using visualization and support tools, which are described below. Next, probes are coupled to the second wheel on the opposite side and second wheel is inspected, too. After evaluation of second wheel, operator decides about wheel condition, while using a straightforward workflow-based inspection procedure and saves data and test results. Finally, underfloor inspection system is moved to the next wheelset to be inspected. A water reservoir with approx. 250-300 liters on board allows to inspect without any interruptions for refill during shift.

At the beginning and the end of each shift inspection system is validated during a Check-/Recheck procure at a test specimen with defined flaws. During Check-/Recheck procedure a failure management system supports operator in validation.

Underfloor inspection system »UFPE-WheelStar« (figure 3) is third generation of underfloor inspection machines and has increased ergonomic conditions and comfort for operator. Therefore, controller and inspection system (in the pit under train) are self-sustaining and linked to each other wireless. Wireless linking not only eliminates risk of broken cables, but also increases comfort for operator. Climbing into pit to fix cables are eliminated. In addition, comfort is increased by a user-friendly workflow-based Human Machine Interface (HMI) for controlling and by user-friendly evaluation software opportunities to validate results. A straightforward workflow-based control concept allows to proceed within logical steps and makes complex inspection task easy to handle. Due to the standardized (workflow-based) proceeding and a high level of automation operators work gets comparable and human factor during inspection is minimized. If there are any doubts during evaluation inspection results can be shared with internal or external experts (e.g. NDT competence center). Experts can support evaluation even from out-of-side or retrospectively. In case of any trouble remote support of OEM is also possible and helps to maximize availability of inspection systems.

Figure 3. Underfloor inspection system UFPE-WheelStar (3rd generation of underfloor inspection system)

### 3. Digitalized in-service wheel inspection

Before discussing 3rd generation of underfloor inspection system in context of industry 4.0 or NDT 4.0, first we have to define industry 4.0 or NDT 4.0 in context of in-service-wheel inspection.

Industry 4.0 is one of the top priorities for many companies and universities, but a generally accepted definition does not exist [2]. Federal Ministry for Economic Affairs and Energy in Germany realized »Plattform Industrie 4.0« to coordinate activities for digitalization of industry in Germany. In this context the ministry defines industry 4.0 as the intelligent networking of machines and processes for industry with the help of information and communication technology [3]. But unfortunately, this definition is not concrete and does not give any concrete suggestions to digitalize products or organizations. Hermann M. et al. discussed in a literature-review six principles of how to realize industry 4.0 scenarios in organizations: virtualization, service orientation, real-time capability, interoperability, decentralization and modularity [2].
Although a testing machine is not an organization, let's discuss about how to use these six design principles for digitalization of NDT. How do we use these six design principles within new generation of underfloor inspection system UFPE-WheelStar and what are the benefits?

### *3.1 Virtualization - One turn for inspection*

Virtualization is what ultrasonic inspection is about: Ultrasonic inspection is used to make defects inside of a specimen (under materials surface or within volume) visible. Here, we use ultrasonic inspection to get information about train wheels condition. This includes condition of rim, tread, inner and external face and wheel disk (see also figure 1). After wheelset is lifted and rotates, probes are coupled automated to the wheel and wheel is going to be inspected. Due to 14 probes, which are used simultaneously (in sum inspection system has 19 probes) and phased-array technology a train wheel is inspected within one turn and in less than 90 seconds. After digitalization A-scan, TD-scans and different projection views are available for evaluation of testing results (figure 4).

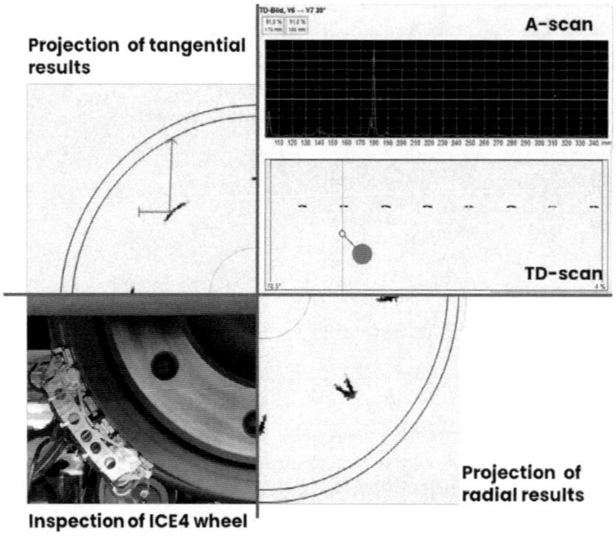


Figure 4. Digitalization of a train wheel within one turn [4]

### *3.2 Service orientation - Straight forward workflow-based inspection procedure*

Due to high inspection speed there are lots of data within short time interval. To keep overview over all data on the on hand and to handle complex inspection system on the other hand, an intuitive straight forward workflow-based inspection procedure is realized and different supporting tools for operator are available.
Support for operator is not only during evaluation necessary, but also during complete inspection process. First, positioning and coupling of probes is fully automated. Thus, positioning and coupling of probes is not affected by human factors and results get comparable. During validation of inspection system (check and recheck to reference testing machine at the beginning or ending of each shift) a specimen-reflector-management pre-evaluates flaws based on an automated comparison of flaw and calibration data. Operator uses pre-evaluation to get information about probe condition and to validate inspection system and inspection results. During evaluation different tools for visualization summarize A-data in projection views or patented difference-method masks circumferential echoes. After evaluation workflow-based inspection procedure guides operator through decision making process as well as creation of protocol and data storage.

In case of any trouble, an integrated condition monitoring helps OEM to support operator remotely – including remote-control (if operator allows access).
Figure 5 summarizes support for the operator during inspection procedure, evaluation and in case of any trouble.

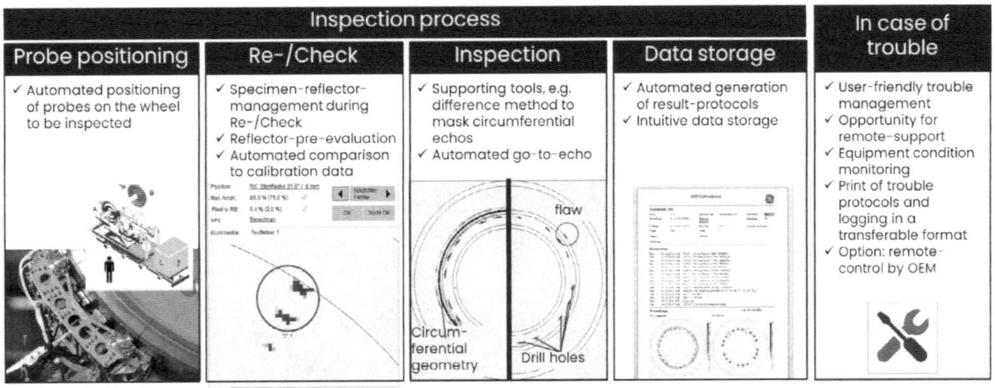


Figure 5. Support during inspection procedure, evaluation and in case of any trouble

Support of inspector during use of inspection system and evaluation of results increases comfort, machine availability and comparability of inspection results and thus increased security, due to human factor is minimized.

### *3.3 Real-time capability – Data management and exchange*

For any kind of support, data mining and data transfer are essential. In concrete relevant data are ultrasonic specific data (e.g. (TCG-curves, calibration and virtual probe gains), comments, protocols, data of condition monitoring, data about operator and of course inspection results.
These data are collected and saved on the inspection system. Data is saved in a for UFPE-WheelStar-inspection systems and stand-alone viewer system compatible, exchangeable format. Due to compatible and exchangeable format, data can be transferred to other UFPE-WheelStar-inspection systems or stand-alone computers (with stand-alone-viewer software installed) or can be stored on serves or in a cloud (depending on operators IT-periphery). Of course, data has to be secured. For example, data relevant for inspection can not only copied and passed from another system but has to be validated by inspection supervisor before first-time use.

### *3.4 Interoperability & Cooperation*

Exchange of UFPE-WheelStar-data increases cooperation and interoperability within company, especially if many UFPE-WheelStar -inspection systems are used and several operators (perhaps placed on different locations) cooperate (figure 6).

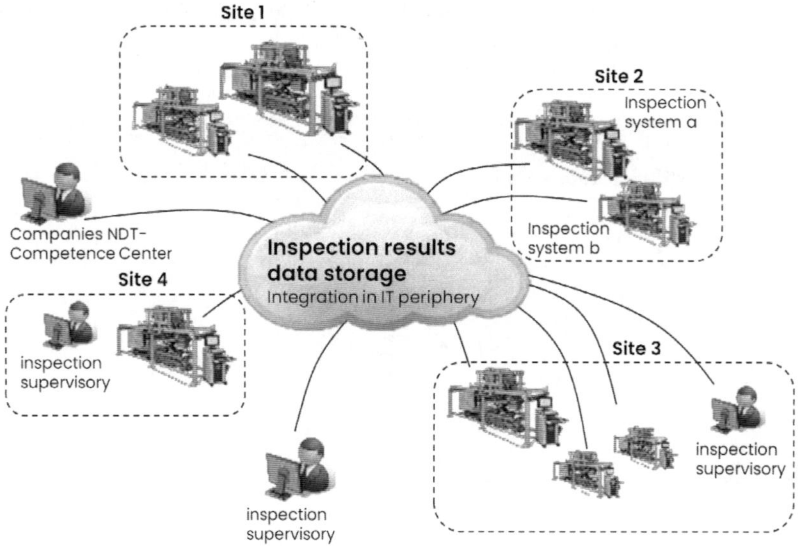


Figure 6. Exchange of UFPE-WheelStar data increases cooperation and interoperability within company

Data or opinion exchange is beneficial during decision making process: In case of any uncertainty from operators point of view, interoperability allows to consult internal or external experts such as inspection supervisory or companies NDT-Competence Center. Due to a stand-alone viewer software and options for data exchange inspection supervisory or companies NDT-Competence Center can support operator during evaluation even if there are not on site. So, decision making process does not depend on operators experience or opinion anymore, but evaluation process gets more comparable. In case of any trouble interoperability also means remote support, mentioned above.

### *3.5 Decentralization - Decision making from everywhere*

These ideas lead to a centralized decision-making process but to a decentralized working surrounding. Know-how and experience are transferred over limits of sites. To do so, standards for data exchange have been developed for UFPE-WheelStar-inspection system.
Next step is to integrate data exchange in operators IT-periphery, respecting IT requirements.

### *3.6 Modularity - One Interface to control & evaluate*

Although inspection task of wheel inspection and evaluation is very complex and an inspection system with a high automatization level is complex, too, operator does not take any notice of this complexity. Within UFPE-WheelStar-inspection system there is only one human machine interface to control evaluation software, machine hardware, ultrasonic electronics, data storage and secure relevant functions (figure 7).

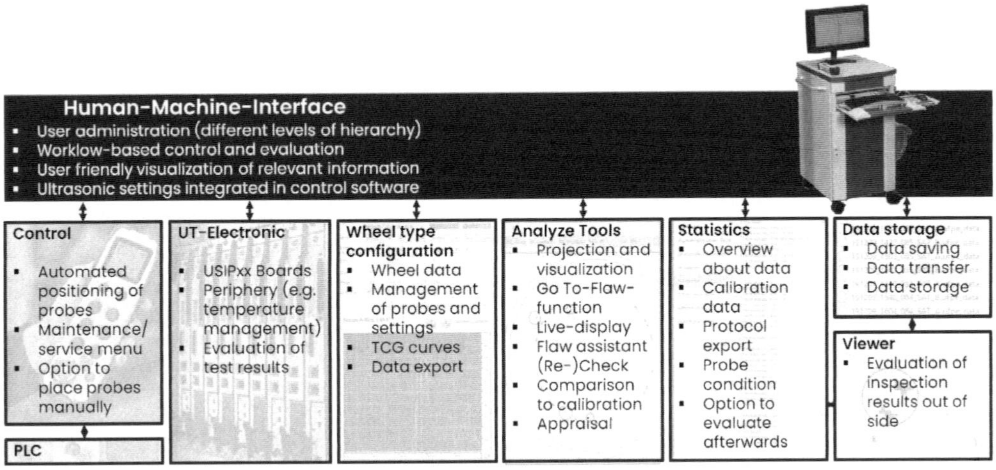


Figure 7. Modular designed hard- and software architecture of UFPE-WheelStar-inspection system

Due to its modular design of hard- and software architecture of UFPE-WheelStar-inspection system, inspection system can be adapted to different wheel types, probe types, test failures, etc. For example, probes can be replaced or added o UT electronic. Protocols and documentation functions can be adapted to customers wish.

## 4. Benefits and conclusion

Due to respecting six scenarios to realize industry 4.0-ideas figured out by Hermann M. et al. [2], UFPE-WheelStar-in-service-wheel inspection system is a good example of how to realize NDT 4.0 in wheel inspection. NDT 4.0 in wheel inspection helps to benefit from digitalization.

Concrete benefits from operators' point of view during inspection procedure are increased comfort during usage. Due to wireless connection and water reservoir time-consuming works within rail track are eliminated and job preparation schedule are reduced. High level of automation and straight forward workflow-based inspection procedure help to reduce causes of error and optimized software opportunities for evaluation and transfer during decision making process, increase comparability of test results and minimize human factor during inspection and evaluation. Operator can concentrate on what is essential: Evaluation of wheels condition.

Operators company benefits due to increased acceptance in face of complex inspection and system technologies, short inspection duration (<90 sec/wheel) and increased system availability due to options for short-term remote support. Thus, UFPE-WheelStar-inspection system is one of the key factors for quality assurance and maintenance of rail.

Many thanks to DB Systemtechnik GmbH and DB Fernverkehr AG.

**References**

1. DIN 27201-7 State of railway vehicles - Basic principles and production technology – Part 7: Non-destructive testing, June 2020, Beuth Verlag, Germany

2. M. Hermann, T. Pentek, B. Otto: Design Principles for Industrie 4.0 Scenarios. A Literature Review. Working Paper No. 01 / 2015. DOI: 10.13140/RG.2.2.29269.22248

3. Federal Ministry for Economic Affairs and Energy Germany, URL https://www.plattform-i40.de/PI40/Navigation/DE/Industrie40/WasIndustrie40/was-ist-industrie-40.html, called 03.07.2020]

4. D. Werner (PhD), A. Franzen, F. Henrix, U. Philipps, P. Buschke: DGZfP Seminar NDT in Railway 2020 in Potsdam, Germany

# Next Generation Tank Floor Scanning – Total Inspection

Stuart G. Kenny[1], Dr Matthew Boat [2]

[1] Eddyfi Technologies

Phone: +44 7775822751; e-mail: e-mail, e-mail, ...
[2] Eddyfi Technologies, Swansea, UK Email: skenny@eddyfi.com

**Abstract**
These instructions have been prepared to assist authors in the preparation of papers for reproduction in Above Ground Storage tanks are subject to obligatory and periodic inspection routines. Due to the harsh operating environments and the catastrophic impact of failure, storage facilities are governed by regulatory practices to ensure the assets are fit for service.

There are numerous potential damage mechanisms associated to the in-service operation of tank storage and each risk of failure requires careful management and routine inspection. Underfloor corrosion is specifically related to the soil interface underneath the tank floor. Regarded as one of the most significant threats in tank storage, floor corrosion often is identified as a risk during integrity management programs and inspection of this potential defect will regularly be included within the written scheme of examination/inspection.

Due to the large service area of a tank floor, the typical inservice condition of the scanning service and the general working conditions inside a storage tank, the inspection method identified to detect underfloor corrosion, must be fast, robust, reliable and provide a high value of confidence for minimum detectability.

Magnetic Flux Leakage systems have been historically deployed for this part of the inspection process and the technology has a proven track record for being able to provide consistent results within these harsh and uncomfortable environments.

However, the historical practices have been challenged for inefficiency and asset owners are insisting for improved productivity, without sacrificing accuracy and traceability.

The latest generation of MFL technology has been developed to significantly improve inspection efficiencies and has identified specific areas within the work flow and scanning parameters that can provide an asset owner with a final report in as minimal time as possible. This results in direct cost savings for reduction in labor time, indirect savings by getting the tank back in-service and reduces risk of confined entry by minimizing the time spent inspecting.

Such developments include; Minimising setup time using software driven inspection plans, onboard LED lighting system to reduce the need for pre-inspection setup, paperless reporting strategy that will include assisted defect recognition and focussed supplementary inspection, minismised dead-zones to maximise floor coverage and curved scanning capabaility to map the critical zone of the annular plates.

Key words: Tank Inspection, MFL, integrity management, minimum detectability

## 1. Introduction

A tank inspection project is made up numerous components, takes precise planning and has significant operational impacts across the site. In addition to the inspection these operations include product management, cleaning activities and potentially building scaffold access. All of these activities have considerable cost to the asset owner and therefore getting the tank back into operation as efficiently as possible is extremely important.

It is the asset owners desire to get the tank cleaned, inspected and reported within the quickest time possible and to have the highest level of confidence in the data output and operational competency.

With the above in consideration, a new age of tank floor scanning is soon to be available. Using the experience of 3 decades of product development and taking advantage of the highest end technology, the new inspection equipment can scan more quickly, detect smaller defects and report more accurately on any areas of damage.

This paper presents key functionality of the new MFL tank scanner to help display the overall efficiency improvements that can make sure a reliable tank inspection report is delivered in the quickest time possible.

## 2. System Overview

### 2.1 Interactive Laser Guide

marking the floor while you map is now fast and easy, thanks to laser-assisted defect location. The laser line is simulated live in the C-scan, allowing to correlate the physical location of tank floor defects with C-scan indications.

Fig 1: laser positioning

This software coordinated laser positioning guide allows the user instantly mark the area of the floor that may require further investigation for reporting. Without pausing the scan or repositioning, the operator will quickly locate the defect on the real time CSCAN, place the coordinated cursor over the defect by moving the scanner head and turn the laser on to mark the spot on the floor.

### 2.3 Onboard Powerful lighting

Front dimmable, focalized LEDs, with diffusing lenses and rear LEDs. Based on chip-on-board technology providing high and uniform intensity, inline with API recommendations for visual inspection, to perfectly highlight product-side pitting and scan obstacles.

Fig 2: Lights

### 2.4 Adjustable bridge height

Reach maximum sensitivity where conditions allow, and easily raise the bridge to scan even in the most challenging conditions, such as undulated plates, repair plates and lap welds or when surface preparation is not ideal.

For optimal MFL performance it is desierable to get the magnet and sensor boards as close to the test surface as possible. This will ensure that the plate is fully saturated and that the smallest defects can therefore be resolved. However, it is clear that any 2 tanks are not identicale and there are many restrictions that can prevent the sensor head getting close to the surface. In the past, operators may have recorded dead areas in the report, or relied on other technology to cover these areas. The new MFL scanner can now mechanically change the height of the bridge and although in some circumstance, sensitivity maybe sacrificed, it will make sure the inspection is completed and coverage is maximised.

### 2.5 SmartMAGNET™

Variable automated magnetic flux strength for optimized inspection performance depending on plate thickness. Besides, literally turn off powerful rare earth permanent magnets for easier storage and shipping, and safer handling.

Possibly the most significant development feature of the new system, the SmartMAGENTTM has a motor driven rotating pole piece that allows the system to be set into 2 flux density positions. On thin floor plates, the low position is pre-set and the operator will benefit from reduced drag and power consumption, however on thicker plates the high flux position is selected and minimum detectability for corrosion defects is improved.

In addition to the performance optimisation, the magnet can also be totally switched off using the same motor drive. This not only improves transportation but cleaning, scanning handling and positioning is now much easier.

Fig 3: Showing the ease of passig the scanning head through a manway

### 2.6 Precision Active Steering

For fast, high-quality curved scans in the critical zone. It drastically reduces dead zones by placing sensors where they are needed most: within 12mm of the tank shell. Comes with the capability to return to straight line driving at the simple push of a button.

The cylindrical shape of a storage tank means that there are inherent deadzones when using a traditional straight line scanning system. It is unfortunate that the most critical part of a storage tank is the area directly adjacent to the tank shell and is often referred to the annular plate section. These plates are generally slightly thicker than the standard floor plates and are regarded as the load bearing section of the floor.

Typically these areas are scanned with supplementary inspection technology as there has not been a system that can efficiently scan straight sections of the floor, but also get into the tight space of the annular curvature.

The new MFL scanner has steering capability that allows the system to scan around a curve as tight as 5 metre/16.5 feet and get the sensors within 10mm of the annular weld. This feature will revolutionise the way that tank inspections are managed and no longer will inspectors need to use UT or other devises to scan the critical zone.

Figure 4 : critical zone scanning

### 2.7 Paperless Reporting

Seamless and intuitive work flow the ensures the operator does not have to pick up a pen throughout the inspection. The custom built tablet and automated defect listing software, means that if an inspector sets up the plate dimensions and scan sequence before inspecting, a pen and paper will not be required. This will reduce the overall inspection time and more importantly improve traceability and confidence in the reported results at the end of the inspection. The software allows for visual inspection and UT data to be imported directly into the report and this information is exported as part of the inspection file at the end of the shift.

## CONCLUSIONS

MFL technology is tried and tested within the tank inspection industry. Its track record for speed, ease of use and reliability of detection has seen it become an integral part of integrity strategy programs. However, the industry is looking for efficiency improvements and does not want to sacrifice output and performance. The unique features identified in this paper are significant in their independence, however combining these improvements into one system has resulted in the total step change the industry has been requesting

# Acoustic anisotropy in products obtained by the method of selective laser melting

Vitaly V. MURAVEV[1,2], Olga V. MURAVEVA[1,2]

[1] Kalashnikov Izhevsk State Technical University, Izhevsk, Russia, e-mail: vmuraviev@mail.ru

[2] Udmurt Federal Research Center, Ural Branch, Russian Academy of Sciences, Izhevsk, Russia, GE,

**Abstract**
The report presents the results of research of acoustic properties of low-carbon 2% Mn-doped steel specimens manufactured by the method of selective laser melting. The electromagnetic-acoustic method is used to input and receive transverse waves of various polarizations, longitudinal and Rayleigh waves. The studies were carried out on samples in the form of flat disks and cylinders. The samples were subjected to different heat treatment to partially relieve internal stresses. The results on non-uniformity of anisotropy over the surface area of disc-shaped samples are presented. The angular indicatrices of the change in the velocity of the transverse tH-wave depending on the direction of polarization and the indicatrix of Poisson's ratio are obtained. The influence of the modes of heat treatment of samples on the velocities of waves of various types, the values of elastic moduli and Poisson's ratio is shown.
**Keywords:** selective laser melting; low-carbon steel with 2% manganese; acoustic properties; velocity; anisotropy; elastic modulus

## 1. Introduction

Additive technologies for creating products are increasingly used in various high-tech industries. Selective laser melting (SLM) is one of the technologies for 3D printing of metals. It is known that the material structure of products obtained by 3D printing technologies using SLM has significant anisotropy of physical and mechanical properties. One of the main problems of SLS is that the residual mechanical stresses appear during the manufacturing process due to the temperature gradient during the local heating and cooling. It's can lead to warping and destruction of products.

A large amount of research is devoted to the study of structure and mechanical properties of these products using different physical techniques. Acoustic methods due to the presence of the most unambiguous relationships between the parameters of elastic waves with measured mechanical characteristics are the most reliable and valid for the assessment of the structural state, anisotropy of properties, and the stress-strain state.

The aim of this work was to experimentally study the acoustic properties of samples manufactured by selective laser melting from the powder of structural low-carbon steel doped with 2% manganese after low temperature annealing and normalization using non-contact EMA method.

## 2. Samples and equipment

The experiments were carried out on samples manufactured by the SLM method in a nitrogen atmosphere from steel powder (0.1% C; 1.5% Mn; <1% Si). Samples were made on an EOSINT 280M SLM printer. The laser beam moved in a zigzag manner within the strip in one direction to the end of the sample, and the strips were rotated relative to the strips of the previous layer by an angle of 15 to 60 degrees. To reduce the temperature gradient, the platform of 3D printer was heated to

100 C. After manufacturing, the samples with the platform were annealed to partially relieve internal stresses at 350 C for 1 (sample #1) and 3 (sample #2) hours. One group of samples was subjected to normalization (annealing at 980 C for 30 minutes, followed by cooling in air) (sample #3).

Samples for acoustic studies were in the form of flat disks 120 mm in diameter and 16 mm thick, with the plane of the disk parallel to the plane of the platform, and the axis of the disk coinciding with the direction of growth of the sample on the 3D printer platform as well as the shape of cylindrical samples with a diameter of 9 mm and a length of 100 mm, with an axis in the plane of the platform.

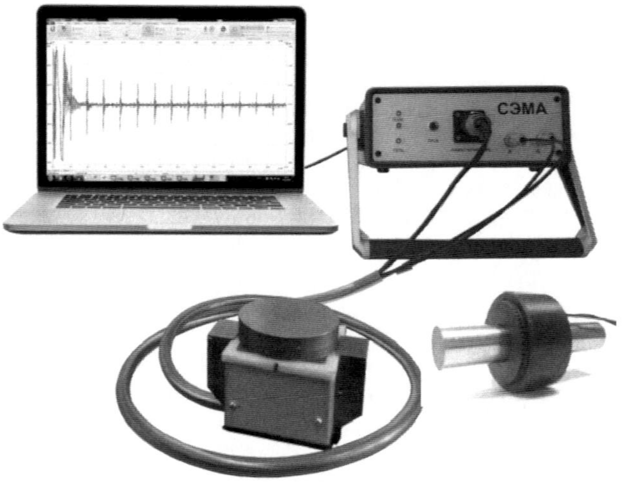


Figure 1 A photo of specialized electromagnetic-acoustic structurescope SEMA

A specialized electromagnetic-acoustic structurescope SEMA was used for acoustic measurements of the samples (Fig. 1). The non-contact electromagnetic-acoustic (EMA) transducer provides the possibility of simultaneous radiation-reception of two transverse tH-waves propagating along the thickness of the sample (growth axis) with mutually perpendicular vectors of horizontal polarization with respect to the plane normal to the growth axis (fig.2a). Fig. 2b shows a series of tH-wave echo pulses detected in two channels and repeatedly reflected from the flat surfaces of the sample. Registration of a series of received signals is performed using specialized software. The observed time delay between the tH-wave pulses polarized in mutually perpendicular directions increased manifold with each subsequent reflection, which indicates the existing anisotropy of acoustic properties in the plane of the disk.

EMA transmission-type transducers for radiation-receipt of acoustic longitudinal and axially polarized transverse waves propagated in all radial directions throughout the objects and surface EMA transducers for radiation-receipt of Rayleigh waves along the bar perimeter (fig.3a). In addition, the propagation time of a longitudinal wave polarized along the growth axis is measured in the same zone using a standard ultrasonic flaw detector DIO 1000PA with piezoelectric transducer with an operating frequency of 10 MHz. The result of measurement is an oscillogram of shear wave multiple reflection series (Fig.3b).

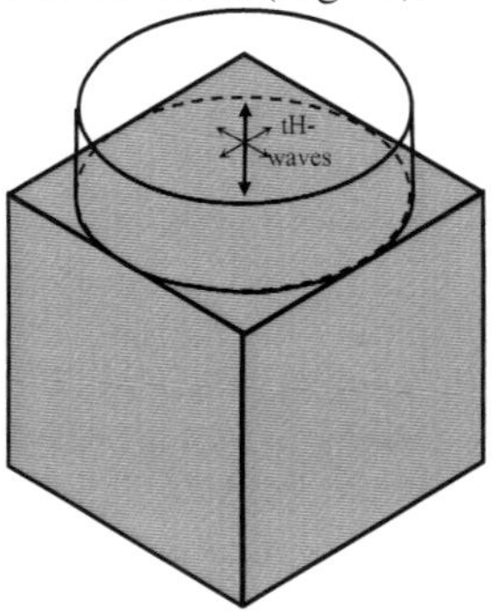


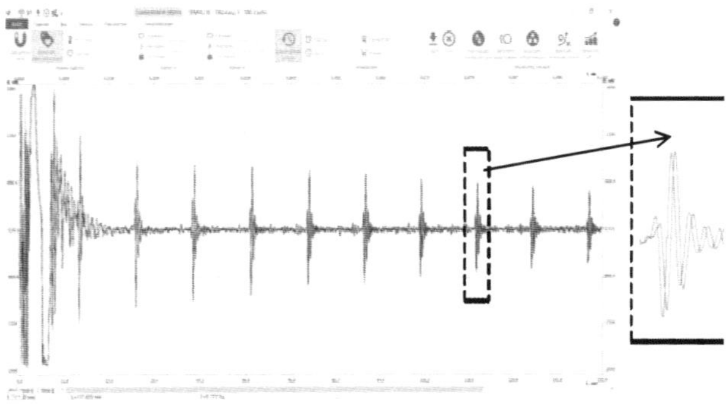

| a | b |
|---|---|

Figure 2 Scheme of propagation of shear tH-wave along the thickness of the sample with mutually perpendicular vectors of horizontal polarization (a), series of bottom echo pulses of two shear waves with orthogonal polarization vectors registered by the SEMA devices (b)

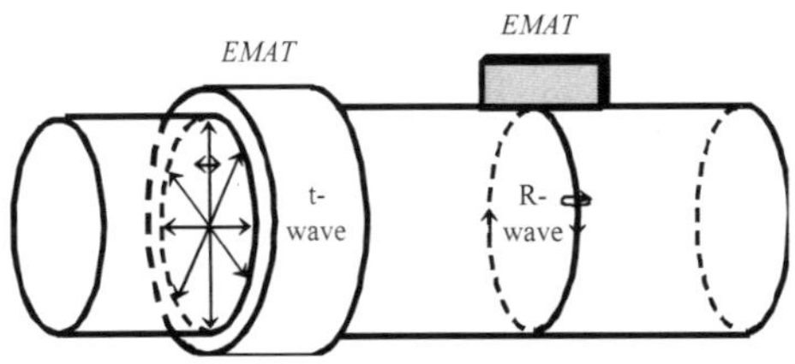


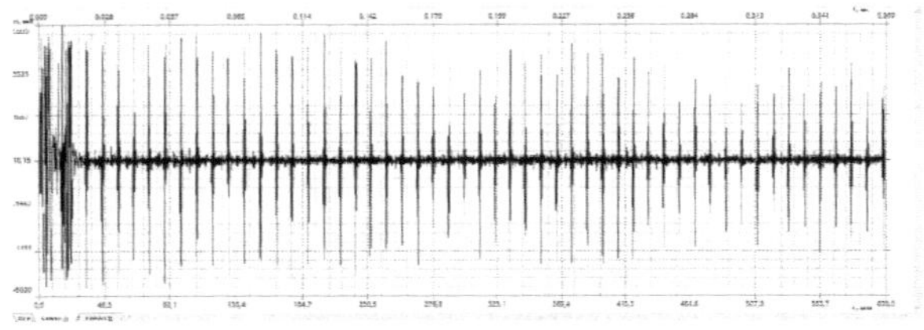

a b

Figure 3 Scheme of propagation of shear t radial waves along the section and Rayleigh R waves along the bar enveloping line (a), specific oscillograms for a series of multiple reflections of the shear wave along the bar diameter (b) series of bottom echo pulses of two shear waves with orthogonal polarization vectors registered by the SEMA device

Velocities of propagation of longitudinal $C_l$, shear $C_t$ and Rayleigh $C_R$ waves are calculated in accordance with formulas:

$$C_t = \frac{D \cdot n}{t_{tn}},\ C_R = \frac{\pi D \cdot n}{t_{Rn}},\ C_{tH} = \frac{H \cdot n}{t_{tHn}},.\ C_l = \frac{H \cdot n}{t_{tHn}}, \qquad (1)$$

where $D$ is the bar diameter, $H$ is the disk thick, $t_n$ is the time of the $n$-th reflection of the pulse in the oscillogram of corresponding wave.

The acoustic anisotropy is determined by the formula:

$$A = \frac{2 \cdot \left[ t_{mn}^{\perp} - t_{mn}^{\parallel} \right]}{t_{mn}^{\perp} + t_{mn}^{\parallel}} \qquad (2)$$

where $t_{mn}^{\perp}$ and $t_{mn}^{\parallel}$ are the delays of the m-th reflected pulses of transverse th-waves, polarized in mutually perpendicular directions, relative to the n-th reflected pulses.

High frequency of discretization of analog-digital conversion, possibility of obtaining a series of multiple reflections and the consequent interpolation provide high accuracy of determining the velocity (0.5 m/s or 0.01%). When assessing the elastic modules of bars (Young modulus $E$, shear modulus $G$, Poisson's factor ν), their functional relation with velocities of propagation of volume (longitudinal and shear) and Rayleigh waves in the object at the known density ρ is used.

## 3. Results and discussion

Fig. 4 illustrates circular diagram in the isolines of the velocity values (velocity indicatrix) relative to the maximum and minimum values of the shear tH-wave. It can be seen that the samples after one-hour and three-hour annealing at a temperature of 350 C (samples # 1 and # 2) retain the anisotropy of properties within ± 4 m / s, while for the samples after normalizing annealing (980 C, 30 minutes) followed by cooling in air (samples # 3), the anisotropy of properties is virtually absent (changes within the error in determining the rate).

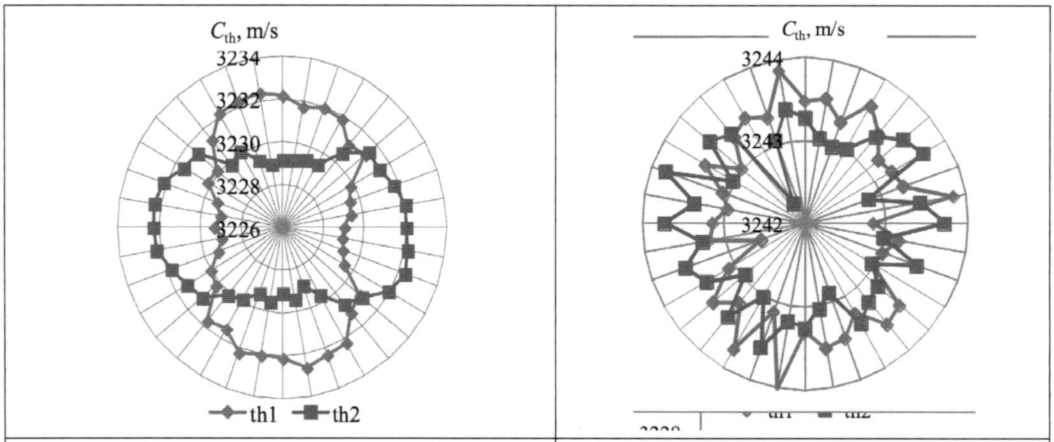


a b

Figure 4 Indicatrix of the velocity of th-waves relative to the maximum value of the velocity (0 degrees) for sample # 2 (a), # 3 (b).

**Table 1. Results of calculation of wave velocities, anisotropy coefficient and elastic moduli**

| Number of samples / quantity | Heat treatment | $C_{tH}^{\|\|}$, m/s | $C_{tH}^{\perp}$, m/s | $C_l$, m/s | $C_R$, m/s | $A$,% | $\bar{\nu}$ | $G$, GPa | $E$, GPa |
|---|---|---|---|---|---|---|---|---|---|
| #1 / 1 | Annealing 350$^0$C, 1 hour | 3231 | 3227 | 5866 | 2978 | 0,265 | 0,2826 | 80,7 | 207,0 |
| #2 / 2 | Annealing 350$^0$C, 3 hour | 3232 | 3228 | 5950 | 2983 | 0,263 | 0,2911 | 80,8 | 208,6 |
| #3 / 2 | Normalization 980 $^0$C, 30 minute, air cooling | 3245 | 3242 | 5953 | 2999 | 0,164 | 0,2900 | 81,4 | 209,8 |

The measured velocities of propagation of th-waves and the calculated coefficient of elastic anisotropy and elastic moduli in the plane perpendicular to the growth direction of the samples are presented in Table 1.

It should be noted that the difference in the longitudinal wave velocities depending on the type of heat treatment is significant, while for the velocities of transverse waves this difference is less significant. According to Table 1, the maximum values of Young's modulus and shear modulus are observed in samples # 3 after normalization, whereas the Poisson's ratio describing the resistance to transverse tensile strain has an average value.

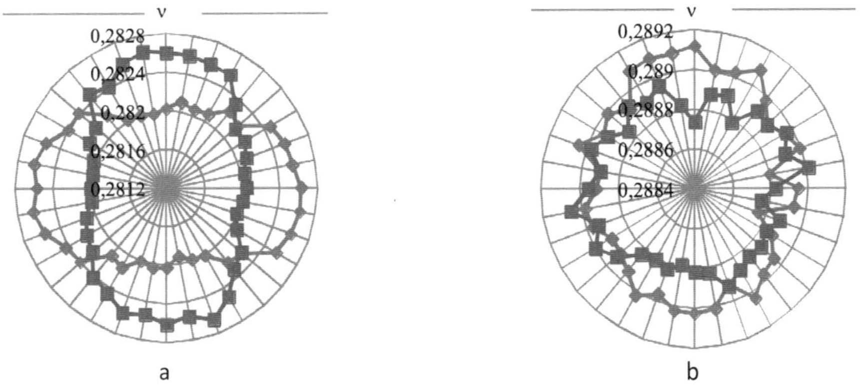


Figure 5. Poisson's ratio indicatrix for samples after different heat treatment in the direction of the maximum value (0 degrees): # 1 (a), # 2 (b)

Fig. 5 shows Poisson's ratio pie charts for the investigated samples. For the samples after low-temperature annealing (# 1 and # 2), the anisotropy of Poisson's ratio is 0.23% and 0.21%, respectively. The anisotropy of Poisson's ratio for sample # 3 after normalization does not exceed 0.1%.

## 3. Conclusions

Thus, the developed by the authors approaches performing the mirror through transmission technique on multiple reflections with use of the EMA transducers provide high accuracy of the main acoustic wave informative parameters evaluation, repeatability and veracity of acoustic structural analysis of the products obtained by the method of selective laser melting. The undertaken studies have shown the opportunity of evaluation of such sensitive-to-structure factors as the velocities of acouctic waves of different types and polarization, anisotropy coefficient and elastic moduli.

**Acknowledgements**
The reported study was funded by Ministry of Science and Higher Education of the Russian Federation (theme “Diagnostics”, No. AAAA-A18-118020690196-3) and by RFBR Bel-a grant, project number 20-58-00015 on the topic: "Investigation of the cyclic durability of steels manufactured by

additive technologies" using the UNU "Information-measuring complex for the study of the acoustic properties of materials and products" (registration number 586308).

# 50 years of conferences DEFEKTOSKOPIE or (un) tolerable lightness of non-destructive testing in Czechoslovakia

# 50let konferencí DEFEKTOSKOPIE aneb (ne)snesitelná lehkost nedestruktivního zkoušení v Československu

Bernard Kopec

Quality Testing Ultrasound

Phone: +49 603 432 965, e-mail: b.kopec@email.cz,
2

**Abstract**
Technical and scientific congresses, conferences or symposia and seminars have been and continue to be important ways of communication in science and technology, whether it is the presentation of innovations and experiences in various fields of human activity or dialogue between the participants in events in corridors and social gatherings. The most important professional and social event of CNDT is the international conference "DEFEKTOSKOPIE", associated with the exhibition of NDT technology and services, which annually becomes a place for exchanging experiences not only CNDT members but also the general public in NDT, welding, construction, energy, manufacturers and pressure equipment operators and others. CNDT tried to maintain a completely uninterrupted tradition of 50 international Defectoscopy conferences. In 2020, the conference was carefully prepared for Přerov, the venue was supposed to be the legendary Hotel Jana, but covid-19 interrupted this uninterrupted tradition. The paper is a historical cross-section of programs and selected related events of all 50 years of "DEFEKTOSKOPIE" held from a personal point of view and preserved memories of the participant of all conferences held so far.

**Keywords:** DEFEKTOSKOPIE, history, non-destructive testing, papers, ČNDT

**Abstrakt**
Technické a vědecké kongresy, konference či sympózia a semináře patřily a patří k významným způsobům komunikace ve vědě a technice, ať již jde o prezentaci novinek a zkušeností v různých oborech lidské činnosti nebo dialogu mezi samotnými účastníky akcí v kuloárech a na společenských setkáních. Nejvýznamnější odbornou i společenskou akcí ČNDT je mezinárodní konference "DEFEKTOSKOPIE", spojená s výstavou NDT techniky a služeb, která se každoročně stává místem pro výměnu zkušeností nejen členů ČNDT, ale i široké odborné veřejnosti z oblasti NDT, svařování, konstrukcí, energetiky, výrobců a provozovatelů tlakových zařízení a dalších. ČNDT se snažila udržet zcela ojedinělou nepřerušenou tradici 50 mezinárodních konferencí Defektoskopie. V roce 2020 byla konference pečlivě připravena do Přerova, místem konání měl být legendární hotel Jana, covid-19 však tuto nepřetržitou tradici přerušil. Příspěvek je historickým průřezem programů a vybraných souvisejících událostí všech dosud konaných 50 ročníků „DEFEKTOSKOPIE" z osobního pohledu a dochovaných vzpomínek účastníka všech dosud konaných konferencí.

**Klíčová slova:** DEFEKTOSKOPIE, historie, nedestruktivní zkoušení, referáty, ČNDT

## 1. Úvod

Historia magistra vitae, Historie je učitelkou života, říká známé úsloví, avšak, jak dodával jeden jezuitský profesor (Giacomo Martina), má velmi málo žáků. „S nadšením studované dějiny totiž mohou a mají mnoho co učit dnešek, který je tolik rozvrácený a žíznící po pravdě, míru a spravedlnosti. Postačilo by, kdybychom se jejich prostřednictvím naučili moudře a odvážně uvažovat o dramatických a zhoubných důsledcích válek, revolucí, pandemií a přírodních katastrof, které poznamenaly cestu člověka po této zemi." Dějiny skrývají svědectví minulosti, která mají být žárlivě střežena a zároveň pomáhat naší cestě do budoucnosti. Dějiny se totiž nemají stávat muzeem nebo hřbitovem nostalgií, nýbrž dědictví minulosti má být dobře zpřítomňováno, jak říká papež František.

## 2. Rok 1970 - rok 1. ročníku DEFEKTOSKOPIE

Projděme se zpátky proudem času roku 1970, rokem konání první konference „DEFEKTOSKOPIE", prvního velkého setkání lidí, zabývající se defektoskopií v Československé republice. Termín „nedestruktivní zkoušení" nebyl ještě v naší republice používán.

Při vzpomínkách na první rok sedmdesátých let minulého století se nedá vyhnout tehdejším událostem, které poznamenaly náš život na dalších 19 let. Rok 1970 asi nebude v českých dějinách zapsán zrovna zlatým písmem. Naším prezidentem byl Ludvík Svoboda, zkušený voják, který bojoval v obou světových válkách, ve skutečnosti byl stát od dubna 1970 pod tvrdým diktátem KSČ v čele s Gustavem Husákem. Čechoslováci zažívali nástup normalizace, měli ještě v živé paměti vpád „spřátelených" vojsk o dva roky předtím. Zemi opustily desetitisíce lidí, dokud nebyly znovu uzavřeny hranice do západních států, namísto někdejšího nadšení se šířily rezignace, apatie, deziluze, deprese a strach. Ve skutečnosti začala hluboká destrukce života v jednotlivých orgánech, organizacích a posléze v celé společnosti. Vyloučení členové strany byli často perzekvováni i různými způsoby, např. výpovědí ze zaměstnání, neumožnění studia jejich dětí aj. My, kteří jsme nebyli členy této strany a tím i nemohli být ve vedoucích funkcích ve všech oblastech života republiky jsme byli „mlčící většinou", a menšina odvážných tzv. „disidenty" (Václav Havel, Václav Malý, Jan Sokol, Tomáš Halík, František Lízna, aj.)

Nebyl to vůbec jednoduchý rok. Svět rozdělovala studená válka, pořád ještě trvala vietnamská válka. Marta Kubišová má definitivní zákaz vystupování a nedostane ani Zlatého slavíka. Paul McCartney oznámil, že odchází z The Beatles z osobních a profesních důvodů (10. dubna 1970) a tím skupina BEATLES, idol naší generace, kterou jsme nazývali spolu s Karlem Krylem, Gottwaldovými vnuky, se rozpadla.

Byly i radostné chvíle v tomto roce. V červnu 1970 se v prachu pařížské antuky zrodil velký tenisový příběh s Janem Kodešem, pozdějším dosud jediným naším wimbledonským vítězem. Zaplněný centrkurt na Roland Garros aplaudoval v neděli 7. června novému šampionovi.

Alexander Solženicyn, nebojácný kritik stalinského teroru a sovětského režimu obdržel Nobelovu cenu za literaturu, o čemž jsme se mohli dozvědět pouze z vysílání Rádia Svobodná Evropa.

Televizní silvestrovské pořady měly nesrovnatelně větší profesionální úroveň, než je tomu nyní. Filmy tzv. „nové vlny" českého filmu měly světovou úroveň. Připomeňme si jeden, dokončený v roce 1970: „Hogo fogo Homolka". Karel Gott nazpíval ty nejlepší hity, jejichž autoři hudby nám byli utajováni, a o nichž jsme se mohli dozvídat pouze na Rádiu Luxemburg. Přes různé obstrukce byly vydávány povídky geniálního Bohumila Hrabala. Milan Kundera, nyní světová mediální hvězda vydal Směšné lásky, znovu vstoupil a vystoupil z KSČ a připravoval se na emigraci.

Proběhla validace jedné z nevýznamnějších staveb tehdejší doby - Nuselského mostu v Praze, na jehož konstrukci bylo použito pramenců pro přepínací výztuž, vyrobených v Železárnách a drátovnách Bohumín. Validace byla provedena zátěžovou zkouškou 66 tanků.

Připomeňme si, že v tomto roce 10. března zemřel František Píšek, odborník na slévárenství a hutnictví, akademik ČSAV (* 24. dubna 1886), který redigoval, dosud v oboru nauky o materiálu, nejrozsáhlejší v české technické literatuře několikadílnou „Nauku o materiálu", kterou pak revidoval Luděk Ptáček. Posledním dílem této edice bylo vydání knihy „Nedestruktivní zkoušení materiálů a konstrukcí" Českou společností pro nedestruktivní testování a nakladatelstvím CERM v Brně v roce 2008, a která v podtitulu nese název „Nauka o materiálu IV" [1].

Pravidelná roční setkání defektoskopických pracovníků s pracovníky základního a aplikovatelného výzkumu nedestruktivních metod (SVÚM Praha-Běchovice) a s výrobci defektoskopických přístrojů (Laboratorní přístroje, n. p. Chotutice) se datují od roku 1966 v Praze. Mezinárodní konference defektoskopie v Československu v těchto létech začala organizovat Angela Houbová z Domu techniky Bratislava v Tatrách (Starý Smokovec). V roce 1970 bylo celostátní setkání pracovníků defektoskopie poprvé nazváno DEFEKTOSKOPIE 1970 a od tohoto roku se začaly počítat ročníky těchto setkání až do dnešního 51. ročníku s malou výjimkou přehlédnutí posloupnosti v roce 1998. [2].

Z lítosti konstatujeme, že jubilejní 50. ročník, připravovaný do hotelu Jana v Přerově se z důvodu pandemie coronaviru nekonal.

**2. Přehled všech dosud konaných konferencí (seminářů) DEFEKTOSKOPIE 1970 - 2019**

**1. ročník DEFEKTOSKOPIE 1970** (nazývaný seminář) se konal ve dnech 17.-19. listopadu 1970 v Domě spojů (Radiopalác) v Praze – Vinohrady. Pořadateli semináře jsou na pozvánce uvedeni Dům techniky Československé vědeckotechnické společnosti (ČSVTS) a Laboratorní přístroje Chotutice, n. p. Iniciátorem a hlavním organizátorem a zakladatelem těchto pravidelných setkání byl pan Mojmír Věchet z Celostátního defektoskopického střediska Státního výzkumného ústavu materiálu (SVÚM) Praha-Běchovice, později zvolen v roce 1990 prvním prezidentem Československé společnosti pro NDT. Semináře se zúčastnilo přes 400 lidí. Bylo předneseno 22 referátů. První referát k 75. výročí defektoskopie prozařováním přednesl Jaroslav Slaba z SVÚM Praha. Na tomto prvním semináři vystoupila řada významných osobností, které tvořily historii české defektoskopie: Dr. Ing. Miroslav Baimler z SPŠJT Praha (RT), Ing. Jaroslav Obraz, CSc. ze SVÚSS Praha (UT), prof. Dr. Ing. Oldřich Taraba, CSc. z ČVUT Praha (UT), Ing. Lubomír Duben, CSc. (MT, ET).

Vynálezce betatronu, urychlovače vysokoenergetického záření pro radiografii velkých tlouštěk, Ing. Karel Rytina z ÚFP ČSAV vystoupil v sekci, zabývající se dnes již málo používanými technikami (neutronová radiografie, mžiková rentgenografie). Na tomto prvním semináři, a i na dalších až do roku 1972, vystoupil zástupce Laboratorních přístrojů Chotutice (Praha-Braník) s informacemi o základním programu vývoje defektoskopických přístrojů. Již na prvním semináři a pak na všech dalších konferencích se uskutečnila výstava defektoskopických přístrojů, tehdy pouze českého výrobce z Chotutic. Je třeba připomenout, že tyto přístroje, především zařízení pro magnetickou metodu práškovou, navazujíce na předválečnou výrobu Ing. Karáska (INKAR) byly srovnatelné s tehdejšími světovými výrobci, To se týkalo i vystavovaných ultrazvukových defektoskopů UID-S a UID-R, jejichž ideovým tvůrcem byl Jaroslav Obraz, pozdější třetí president ČNDT. Tehdejší vedoucí Celostátního defektoskopického střediska při SVÚM Praha Ing. Lubomír Čížek, CSc., informoval o 6. mezinárodní konferenci NDT, konané v roce 1970 v Hannoveru. Mezinárodní konference NDT byly tehdy pořádány nepravidelně ve tří až čtyřletých intervalech po patronací International Committee for NDT (ICNDT), který byl ustaven v roce 1960 a pak i pod patronací European Federation for NDT(ECNDT), který byl založen v roce 1998. První, doložitelná mezinárodní conference NDT se konala v Bruselu v roce 1955. Skutečný zájem tehdejší československé defektoskopické veřejnosti byla nejen účast více než 400 účastníků prvního ročníku DEFEKTOSKOPIE, ale i 6 příspěvků na světové konferenci v Hannoveru z Československa. Přednášejícími v Hannoveru byly významné osobnosti, tvořící dějiny NDT: Josef Krautkrämer (UT), Robert Halmshaw (RT) a Friedrich Förster (ET). První konferenci uzavřel referátem o úsměvných zážitcích dvouletého působení zavádění defektoskopie v Indii Mojmír Věchet.

**2. ročník DEFEKTOSKOPIE 1971** se konal ve dnech 8,-10.listopadu 1971 rovněž v sále Domu spojů (Radiopaláci) v Praze, naposledy v tomto místě stejnými pořadateli. Bylo předneseno 22 referátů, účast rovněž velká, přes 400 účastníků. S referátem o použití měrek v radiografii vystoupil Ing. Evžen Honců, CSc., jehož přednášky i na dalších konferencích se vyznačovaly vytříbeným stylem. Jeho psané referáty považuji za vzor stylistiky v technické literatuře NDT. Poprvé vystoupil Ing. Richard Regazzo (SVÚM) referátem z oblasti zkoušení svarů, člověk, který posvětil svůj život plně ultrazvukové defektoskopii. Poprvé se na konferenci prezentoval Ing. Václav Bízek, CSc. (SVÚM), významný odborník v oboru radiografie. Ředitel LP Chotutice a jeho spolupracovníci prezentovali nadějný výhled výroby defektoskopických přístrojů v Československu. Naděje tohoto výhledu pohasla brzy v roce 1973 zastavením výroby dle rozhodnutí Státní plánovací komise.

**3. ročník DEFEKTOSKOPIE 1972** se konal ve dnech 12.-13. prosince 1972 poprvé v Raisově sále Ústředního domu železničářů (Národním domě) v Praze, účast 450 lidí. V záhlaví pořadatelů se objevila Odborná skupina pro defektoskopii ČSVTS a CDS Praha. Za připomenutí stojí první referát významného odborníka defektoskopie jaderných reaktorů Ing. Jiřího Přepechala, CSc. Poprvé se prezentoval doc. Ing. Rudolf Dubenský, CSc., pozdější druhý prezident ČNDT. Zazněly referáty dvou významných praktiků ze Slezska a to Ing. Vladimíra Matušky z Třince (tehdy Železárny V. M. Molotova) a Ing. Jiřího Ullmana z Kunčic (tehdy Nová huť K. Gottwalda), jehož referáty a články v odborném periodiku Hutník, by mohly být vzorem pro tvorbu srozumitelných prezentací.

**4.ročník DEFEKTOSKOPIE 1973** se konal rovněž v Raisově sále Ústředního domu železničářů (Národním domě) v Praze, účast nižší, přesto 300 účastníků. Poprvé se představili příspěvky další osobností: Ing. Jiří Blahůšek, CSc. (SVÚM), průkopník české normalizace v defektoskopii a doc. Ing. Miroslav Samotný, CSc. (VUT Brno), odborník metody magnetické metody práškové.

**5.ročník DEFEKTOSKOPIE 1974** se konal rovněž v Raisově sále Ústředního domu železničářů (Národním domě) v Praze, účast stále vysoká 350 lidí. Seminář se konal pod záštitou 25 let Státního výzkumného ústavu materiálu (SVÚM), pod kterého spadalo Celostátní defektoskopické středisko (CDS). Sluší se připomenout vystoupení odborníka z oblasti radioisotopů Ing. Zdeňka Frynty (SVÚM).

**6.ročník DEFEKTOSKOPIE 1975** se konal ve dnech 11.-12.11 1975 v Hajnovce v Praze – Vinohrady, účast 300 lidí. Zastavením výroby defektoskopických přístrojů v Laboratorních přístrojích Chotutice, se poprvé prezentovali firmy zahraničních výrobců prostřednictvím českých zastoupení, které ovládly trh.

**7.ročník DEFEKTOSKOPIE 1976** se konal ve dnech 15.-18.11 1976 poprvé mimo Prahu, a to v Beskydech v hotelu Visalaje v Krásné a to především zásluhou významné osobnosti historie české defektoskopie Ing. Leoše Kláska, vedoucího defektoskopického střediska tehdejších Vítkovických železáren (VŽKG) pozdějšího iniciátora založení ČNDT. Prezentovali se zde zástupci tehdejšího nejvýznamnější defektoskopického pracoviště v Československu (Ing. Leoš Klásek, Josef Solnař, Ing. Stanislav Janovský, Ing. Petr Šípek, Ing. Olga Ličková).

**8.ročník DEFEKTOSKOPIE 1977** se konal ve dnech 1.-2.10. 1977 se konal rovněž mimo Prahu, a to v hotelu Panoráma v Mariánských Lázních, v době po vydání Charty 77. Na semináři se objevil první referát z akustické emise pod názvem Využití metody akustické emise pro sledování růstu trhliny ve svarovém spoji při statickém zatěžování autorů Jiřího Blahůška (SVÚM), Václava Kouly (SVÚM) a Františka Čermáka (ČVUT Praha). Poprvé byla přednesena přednáška z oblasti zkoušení těsnosti při výrobě součástí a celků jaderných elektráren významného odborníka Ing. Františka Uhra (Škoda ZVJE Plzeň).

**9.ročník DEFEKTOSKOPIE 1978** se konal ve dnech 25.-28.9.1978 v hotelu Vsacan ve Vsetíně. Zde se poprvé prezentoval odborník defektoskopie vířivých proudů Ing. Bohumil Brodský (VTŽ Chomutov). Celá jedna sekce byla věnována defektoskopii v jaderné elektrárně Jaslovské Bohunice (VÚJE).

**10.ročník DEFEKTOSKOPIE 1979** se konal ve dnech 29.10.-1.11.1979 podruhé v Mariánských Lázních v hotelu Casino. Byl uspořádán jako součást 30 let SVÚM Praha. Poprvé vystoupil Prof. Ing. Bohumil Míšek (VUT Brno), organizátor tří běhů postgraduálního studia defektoskopie a jedna z nejvýznamnějších postav historie NDT. Druhý seminář, ve kterém se objevila akustická emise, poprvé vystoupili lidé, kteří dodnes tvoří historii NDT Ing, Václav Svoboda (Sigma Modřany) a Ing. Zdeněk Převorovský (Ústav thermodynamiky ČSAV Praha).

**11.ročník DEFEKTOSKOPIE 1980** se konal ve dnech 11. – 13.10.1980 poprvé v Hradci Králové. Ve sborníku referátu z konference jsem objevil poznámky z diskuze Jaroslava Obraze, Luďka Ptáčka, Bohumila Míška k referátům Ing. Richarda Regazza. Diskuse v těchto ročnících konferencí měly vysokou pracovní a odbornou úroveň, byly někdy dramatické, ale vždy etické.

**12.ročník DEFEKTOSKOPIE 1981** se konal ve dnech 6.-8.10.1981 potřetí v Mariánských Lázních v hotelu Casino. Zde stojí za zmínku sekce magnetických metod s referáty Miroslava Samotného, Jiřího Šáchy, Josefa Solnaře.

**13.ročník DEFEKTOSKOPIE 1982** se konal po několika létech Praze. Byl zajímavý tím, že byly prezentovány referáty z tehdy se rozvíjející techniky neutronové defektoskopie. Z oblasti rozšiřujícího využití defektoskopie ve výrobě železničního dvojkolí na evropských železnicích byl prezentován referát Bernarda Kopce a Michala Zdukose (tehdejších Železáren a drátoven Bohumín).

**14.ročník DEFEKTOSKOPIE 1983** se konal v Praze. V tomto období již byla československá defektoskopie odkázána na dovoz defektoskopických přístrojů prostřednictvím českých zástupců. Na této konferenci prezentoval Ing. Jiří Schmied nový výrobní program, tehdy nejvýznamnějšího výrobce ultrazvukových defektoskopických přístrojů firmy Krautkrämer. Za zmínku stojí, že některé z přístrojů tehdy prezentovaných jsou využívány dodnes. Ing. Jiří Schmied, respektovaný odborníky firmy Krautkrämer má nemalou zásluhu na efektivnosti české ultrazvukové defektoskopie.

**15.ročník DEFEKTOSKOPIE 1984** se konal v Praze ve znamení 35 let SVÚM. V tomto roce zemřeli RNDr. Jaroslav Slaba, prof. Ing. Jan Prudký a Mirolav Tryta, lidé, kteří posunuli vývoj české defektoskopie.

**16.ročník DEFEKTOSKOPIE 1985** se konal ve Žďáru nad Sázavou. Na tento seminář mám osobní vzpomínku, kdy Mojmír Věchet mi přenechal řízení plenární sekce prvního dne.

**17.ročník DEFEKTOSKOPIE 1986** se konal v Karlových Varech, Poprvé vystoupili zahraniční účastníci, a to hned z několika států: Německa (tehdy NSR), tehdejší Jugoslávie, Bulharska, Rakouska, Maďarska, tehdejšího Sovětského svazu a Rumunska s 12 referáty. Ralph Giese (NSR) prezentoval počítačově řízené rentgenové zařízení firmy Richard Seifert a Günther Gausse prezentoval nová magnetická zařízení firmy Tiede. Poprvé vystoupila mezinárodně uznávaná osobnost NDT Rainer Frielinghaus, vedoucí aplikační laboratoře firmy Krautkrämer.

**18.ročník DEFEKTOSKOPIE 1987** se konal ve dnech 3.-5.10.1987 ve Znojmě. Za zmínku stojí referát k přínosu elektroniky k ovládání magnetizačních přístrojů Mojmíra Věcheta. Zahraniční účastníci nebyli.

**19.ročník DEFEKTOSKOPIE 1988** se konal poprvé v Táboře. Za zmínku stojí vystoupení Ing. Zdeno Figury z významného slovenského ultrazvukového pracoviště Výzkumného ústavu mechanizácie a automatizácie Nové Mesto nad Váhom. Ve vinárně na táborském náměstí byl iniciován vznik ČNDT (L. Klásek, zemřel 1991, M. Věchet, B. Kopec).

**20.ročník DEFEKTOSKOPIE 1989** se konal ve Znojmě. Bylo to těsně po sametové revoluci. Bylo připomenuto 50 let vzniku SVÚM. Za připomenutí stojí referát jednoho z mála odborníků NDT na železnici Ing. Vlastimila Buriana (Železniční opravny Nymburk) k využití defektoskopie v železničním opravárenství.

**21.ročník DEFEKTOSKOPIE 1990** se konal podruhé v Ostravě. Sborník referátů měl poprvé a naposled 2 díly. Prezentováno bylo rekordních 45 referátů. Podruhé se zúčastnili zahraniční hosté, a to z Německa, Rakouska, Bulharska, Sovětského svazu a Maďarska. Referátem se prezentovala další světová osobnost ultrazvukové defektoskopie Udo Schelngermann, současný předseda normalizační komise ISO SC 3 pro NDT zkoušení ultrazvukem z firmy Krautkrämer.Byl to první seminář po ustavujícím sjezdu ČNDT, který se konal na podzim roku 1990 v Kuřimi, na kterém bylo zvoleno první předsednictvo: prezident Mojmír Věchet a dva viceprezidenti: za Moravu Ing. Jiří Ullman, za Slovensko Ing. Pavol Gabriška.

**22.ročník DEFEKTOSKOPIE 1991** se konal v říjnu, poprvé ve Zlíně. Poprvé byla jako pořadatel uvedena Československá společnost pro nedestruktivní testování. Poprvé byla rovněž prezentována přednáška o nově budovaném systému kvalifikace a certifikace pracovníků NDT v ČSFR Ing. Jiřího Pittera, navazujíce na klasifikaci CDS, kterou na předcházejících seminářích prezentoval systematicky Oldřich Miffek.

**23.ročník DEFEKTOSKOPIE 1992** se vrátil v říjnu do Tábora. Byla rozběhnuta komercializace československé defektoskopie, bylo založeno několik firem, jejichž vlastníci se vyčlenili z bývalých státních podniků, kapitalismus v defektoskopii začal velice brzy po sametové revoluci se všemi klady i zápory. Ve sborníku se poprvé objevily reklamy nových podniků. Byl zvolen nový prezident ČNDT doc. Ing. Rudolf Dubenský, CSc.

**24.ročník DEFEKTOSKOPIE 1993** se konal v říjnu v Táboře. Za zmínku stojí referát jednoho z nevýznamnějších odborníků v metodě netěsnosti Ing. Petera Žubora z Piešťan k výchově pracovníků metod netěsnosti.

**25.ročník DEFEKTOSKOPIE 1994** se konal v říjnu počtvrté v Táboře. Poprvé se objevili sponzoři uspořádání konference: Olympus, Testima. ATG, TS Systém, Helling, SVÚM, PAPCo., Defekta, Balteau, BVD. U příležitosti konference poprvé byl poprvé vydán NDT Bulletin.

**26.ročník DEFEKTOSKOPIE 1995** se konal v říjnu popáté v Táboře. Potřetí se zúčastnili zahraniční přednášející a oba s referáty tehdy intenzívně se rozbíhajících systémů klasifikace a certifikace pracovníků, a to Charles Hellier z USA a Dietmar Hennig z Německa. Ve sborníku se poprvé a naposled objevily fotografie přednášejících. Byl zvolen nový prezident ČNDT Ing. Jaroslav Obraz, CSc., jedna z největších postav celé historie československé defektoskopie, uznávaná ve světě.

**27.ročník DEFEKTOSKOPIE 1996** se vrátil do Prahy. Uvažované pokračování místa konání v Táboře nebylo možné z důvodu nevyhovujícího ubytování. Významná účast zahraničních hostů z Německa a Ruska.

28.ročník DEFEKTOSKOPIE 1997 se konal v Praze pod záštitou ministra průmyslu a obchodu. Poprvé se v tiráži sborníku referátu z konference objevilo č.28, V tomto roce začala převaha referátů (většinou kolektivy min. 4 autorů) z výzkumu a aplikací metody akustické emise akademických pracovišť z Brna, která trvá dodnes, kdy počet referátů z této metody převyšuje počty referátů klasických metod. Poprvé se na konferenci prezentoval doc. Ing. Pavel Mazal, CSc, pozdější dlouholetý prezident ČNDT.

***(29.ročník)*** **DEFEKTOSKOPIE 1998** se konal v Praze pod záštitou předsedy ÚNMZ. Nejmenší prezentovaný počet referátů a tím i nejútlejší sborník. Pořadové číslo conference 29 bylo opomenuto.

**29.ročník DEFEKTOSKOPIE** 1999 se konal ve dnech 17. – 19.11.1999 poprvé v Hradci Králové. Zde došlo k chybě číslování konferencí: na obálce sborníku se objevilo č. 29, správně mělo být č.30. Ve sborníku byl poprvé vytištěn rejstřík autorů. Objevila se v programu první tzv. pozvaná přednáška Gerharda Dobmanna z Frauenhöffrerova Institutu ze Sarbrückenu, Byl prezentován jeden referát, pod kterým bylo uvedeno 15 autorů ze 7 států z oblasti ET, přednesený J. Pavó z Maďarska. U příležitosti 50 let existence firmy Krautkrämer, přednesl W. Berke, vedoucí prvního kvalifikačního pracoviště UT v Evropě před vydáním EN 473 (v roce 1984 podepsal certifikát UT autorovi tohoto článku po absolvovaném školení), Přehled o vývoji digitální technologie měření ultrazvukem.

**30.ročník DEFEKTOSKOPIE 2000** se konal ve dnech 21.-23.11.2000 podruhé v Hradci Králové. Do programu byly zařazeny 3 pozvané přednášky (U. Schengermann, W. A. K. Deutsch). Na sjezdu CNDT byl zvolen prezidentem doc. Ing. Pavel Mazal, CSc. a tím došlo k postupnému zefektivnění činností ČNDT a pozvednutí její mezinárodní prestiže.

**31.ročník DEFEKTOSKOPIE 2001** se konal ve dnech 20.-22.11.2001 v Praze. V programu zazněl prognostický referát autorů Luďka Ptáčka, Bohumila Míška a Pavla Mazala: Metody a zkušeností NDT pro 21 století. Zúčastnilo se 9 přednášejících ze zahraničí. Začal převažovat počet referátů z metody AE (tento rok 15).

**32.ročník DEFEKTOSKOPIE 2002** se konal ve dnech19.-21. 11.2002 v Liberci. Poprvé vystoupil na konferenci prof. RNDr. Jaroslav Fiala, CSc., jehož přednášky nepostrádaly srozumitelnost a noblesu vystoupení. Poprvé byly prezentovány referáty z Asie (Malajasie), a z Ameriky (USA).

**33.ročník DEFEKTOSKOPIE 2003** se konal ve dnech 19.-21.11.2003 potřetí v Ostravě. Poprvé vystoupil, pak dlouholetý účastník konferencí, Ing. Vladimír Husárek (Sofranel) z Francie, jehož zkoušení NDT fleretů se objevilo i na olympijských hrách.

**34.ročník DEFEKTOSKOPIE 2004** se konal ve dnech 3.-5.11.2004 ve Špindlerově Mlýně v hotelu Harmony. Podruhé se zúčastnili přednášející z Asie, a to z Iráku. Součástí konferencí byl od vzniku ČSNDT i slavnostní večer. Na tomto večeru se uskutečnilo poprvé a naposled setkání všech prezidentů ČNDT, kterými byli Mojmír Věchet, Rudolf Dubenský, Jaroslav Obraz a Pavel Mazal.

**35.ročník DEFEKTOSKOPIE 2005** se konal ve dnech 8.-10.11.2005 ve Znojmě, Na konferenci se poprvé prezentoval referátem zástupce dalšího kontinentu Afriky, a to z Alžíru. Sluší se připomenout košt a ochutnávka vína v Louckém klášteře z proslulé znojemské vinice Šobes v rámci pravidelně uspořádaných společenských večerů konferencí.

**36.ročník DEFEKTOSKOPIE 2006** se konal ve dnech 7.-9.11.2006 a vrátil se do Tábora. Na konferenci se prezentovalo 6 přednášejících z Polska. Polští odborníci se pak stali pravidelnými účastníky a vystupovali na konferencích několikrát, jako Krysztow Schabowicz (25 příspěvků), J. Hola (12 příspěvků), T.Gozelczak (11 příspěvků).

**37.ročník DEFEKTOSKOPIE 2007** se konal ve dnech 7.-9.11.2007 v Praze. Referáty z metody akustické emise převažovaly. Zajímavostí byl referát přednesený prof. Luboše Pazdery prezentovaný s uvedením osmi spoluautorů.

**38.ročník DEFEKTOSKOPIE 2008** se konal ve dnech 4.-6.11. 2008 poprvé v Brně. Zajímavostí této konference byla prezentace devíti zahraničních příspěvků z Asie (Korea, Irán) a Afriky (Egypt). V průběhu slavnostního večera se uskutečnil křest dlouho očekávané odborné učebnice NDT Nedestruktivní zkoušení materiálu a konstrukcí (B. Kopec a kol. autorů), vydaného ČNDT a nakladatelstvím CERM.

**39.ročník DEFEKTOSKOPIE 2009** se konal ve dnech 4.-6.11.2009 v Praze. Za připomenutí stojí italský příspěvek z muzea v Calabrii o využití NDT ve výtvarném umění a indický příspěvek k využití AE při zkoušení deformace hliníku.

**40.ročník DEFEKTOSKOPIE 2010** se konal ve dnech 10.-12.11.2010 poprvé v Plzni. Je třeba se zmínit o vystoupení A. A. Dubova k fyzikálním základům a využití metody magnetické paměti materiálů (MMM) a prezentací zástupce firmy Honeywell k využití Lambových vln v letectví. Na konferenci a pak i v zasedání odborné skupiny pro technickou normalizaci NDT TNK 80, která od roku 2008 se stala pravidelnou součástí programu, se rozproudila bohatá diskuze k terminologii indikací defektoskopických zjištění iniciována referátem „Co skutečně zjišťuje NDT" (B. Kopec) a lektorskou činností odborného experta ÚNMZ, PhDr. Věry Vlkové, CSc. Exkurze do Plzeňského pivovaru s prezentací jeho historie rozšířilo obzor kulturních znalostí českého vynálezu pivního ležáku.

**41.ročník DEFEKTOSKOPIE 2011** se konal ve dnech 9.-11.11.2011 počtvrté v Ostravě. Za zmínku stojí prezentace autorů L. R. Lindmooda a J. B. Spicera z John Hopkins University, USA, jednoho z nejvýznamnějších světových výzkumných ultrazvukových pracovišť.

**42.ročník DEFEKTOSKOPIE 2012** se konal ve dnech 30.10.-1.11.2012 v Seči u Chrudimi. Na konferenci bylo zaznamenáno nejpočetnější zastoupení polských odborníků (12 prezantací) včetně Piotra Šullika, bývalého člena redakční rady NDT Bulletinu a překladatele publikací J. Obraze do polštiny. Objevil se referát z Turecka. Poprvé se prezentoval souhrnným referátem jeden z mála významných odborníků infračervené thermografie Ing. Jiří Svoboda. V kuloárech bylo diskutováno vytvoření dalšího certifikačního orgánu z iniciativy poskytovatelů školení a zkoušení NDT a monopolního chování APC. Historii kvalifikace a certifikace pracovníků NDT od počátků, vytvořených v Celostátním defektoskopickém středisku, přes vytvoření APC a pak dalších 5 certifikačních orgánu se tento příspěvek z důvodu obsáhlosti problematiky nezabývá.

**43.ročník DEFEKTOSKOPIE 2013** se konal ve dnech 5.-7.11.2013 v Olomouci. Na konferenci byl prezentován mezinárodní projekt PROQUALINDT (P. Mazal, B. Kopec, E. Romero) na rozšíření a sjednocení učebních textů metod NDT, kterého se zúčastnilo 16 společností NDT Evropy.

**44.ročník DEFEKTOSKOPIE 2014** se konal ve dnech 9.-11.11.2014 v Praze v rámci 11. Evropské konference NDT(ECNDT). Získání pořadatelství této nejvýznamnější akce z oboru NDT v České republice, složení a počet účastníků, průběh a ohlas v oboru NDT, získaly české defektoskopii, dosud nedoceněný světový ohlas a prestiž. Největší podíl na přidělení uspořádání této prestižní světové akce a její efektivní průběh má dlouholetý prezident ČNDT doc. Ing. Pavel Mazal, CSc.

**45.ročník DEFEKTOSKOPIE 2015** se konal ve dnech 3.-5.11.2015 podruhé v Brně. Začala série referátů z historie NDT (B. Kopec), pokračující do roku 2019. které připomínají nevýznamnější svědectví z objevů fyzikální základů NDT.

**46.ročník DEFEKTOSKOPIE 2016** se konal ve dnech 8.-10.11.2016 v Luhačovicích. Přesto, že konference se konala v dosud nejkrásnějším prostředí, byla účast poněkud menší. Byl přednesen poprvé referát z Jižní Ameriky (Brazílie).

**47.ročník DEFEKTOSKOPIE 2017** se konal ve dnech 7.-9.11.2017 v Chomutově. Mimořádně vysoká účast, 36 referátů. Návštěva soukromého pivovaru Karásek a Stülpner byla přínosem milovníkům piva.

**48.ročník DEFEKTOSKOPIE 2018** se konal ve dnech 6.-8.11.2018 v Praze. Evidován historicky nejvyšší počet přihlášených referátů (41, z toho 12 zahraničních). K zajímavostem patří referát 9 autorů (mezi jinými Zdeněk Převorovský) ze 4 států z využití metody pro neinvazivní hodnocení slitin v lékařské protetice, zpracovaný v rámci grantu ministerstva v Rumunsku.

**49.ročník DEFEKTOSKOPIE 2019** se konal ve dnech 7.-9.11.2019 poprvé v Českých Budějovicích. Po 15 létech úspěšného a efektivního působení P. Mazala ve funkci prezidenta byl zvolen nový, v pořadí 5. prezident ČNDT doc. Mgr. Libor Topolář, Ph.D. Činnost ČNDT začalo být ovlivněna nešťastnou epidemii coronaviru od začátku roku 2020 do dnešních dnů (říjen 2021). Jubilejní 50. ročník (dle oficiálních doložitelných) psaných záznamů, který byl pečlivě připravován na dny 10.-12.11.2020 v hotelu Jana v Přerově musel být kvůli pandemií coronaviru zrušen.

**Závěr**

Dějiny tvoříme neustále my všichni od nepaměti. Jsou o tom, co se událo, ale i naše přítomnost se stává během vteřiny minulostí a tedy dějinami. Tedy i 50 let historie DEFEKTOSKOPIE. Dějiny tvoří historii, slouží k poučení z minulosti do budoucnosti. Historie umožňuje zkoumat současnost na základě minulosti, bez znalosti minulosti nelze pochopit přítomnost.Ten, kdo se nepoučí z minulosti je odsouzen k jejímu opakování. Citát jehož autorem je George Santayana, což byl americký filosof, kritik a spisovatel španělského původu, představitel naturalismu a „klasické" americké filosofie. Tolik klasik.Navazujíce na název příspěvku, dle mého názoru, který nikomu nevnucuji, představa, že se můžeme poučit z historie, je iluze. Historie není žádná matka moudrosti, ani učitelka života, pravdy vůbec ne. Lidstvo je nepoučitelné a popravdě řečeno, ani to nejde. Respektive to poučení, které bychom si měli vzít je jednoduché: všechny scénáře jsou nedopsané, historie ani budoucnost není nikdy černobílá, vždycky je to takové šedivé, „učmouhané". A co bude, nevíme.Jsme schopni se poučit z historie? Dokážeme si předávat z generace na generaci různá varování? Jednotlivci si pamatují události až desítky let staré, a to někdy i do pozoruhodných detailů [3]. Německý učenec Hermann Ebbinghaus, který ve své práci publikované roku 1885 popsal, že křivka zapomínání má přibližně tvar hyperboly – čím více času uplynulo od vjemu, tím méně si pamatujeme [4]. Chceme-li tedy nějakou informaci dlouhodobě udržet v paměti, neobejdeme se bez opakování.

Epidemie coronaviru změnila běh mnoha událostí ve světě i v našich životech. Doba, kterou prožíváme, zpochybnila naše mnohé jistoty. Připadáme si chudší a křehčí, protože jsme prožili pocit omezení a restrikci svobod. Ztráta zaměstnání, ztráta nejbližších lidí, absence běžných mezilidských vztahů – to vše nám náhle otevřelo obzory, které jsme už přestali vnímat. Otřáslo se naše duchovní a materiální bohatství, uvědomili jsme si, že máme strach.Buďme osudu vděčni, že máme DEFEKTOSKOPII. Že máme možnost diskutovat otázky jednak čistě odborné ze všech oblastí nedestruktivního zkoušení a jednak i se setkat s kolegy, s nimiž nemáme možnost se často vidět. Tato osobní stránka každé konference, tyto kuloárové rozhovory jsou stejně přínosné jako plenární zasedání a jednání v sekcích.

**Literatura**

1. Kopec B. a kol. Nedestruktivní zkoušení materiálu a konstrukcí, CERM, ČNDT, Brno 2008.

2. Sborníky referátů z konferencí DEFEKTOSKOPIE 1970 až 2019.

3. Larzabal C., Bacon-Macé N., Muratot S., Thorpe S. J.: Waking Up Buried Memories of Old TV Programs. Front. Behav. Neurosci. 11, 2017

4. Ebbinghaus H. Memory: A Contribution to Experimental Psychology. Teachers College, Columbia University, New York 1885.

**Tab. 1 Přehled autorů a počtu referátů na konferencích DEFEKTOSKOPIE 1970 až 20019**

**(bez referátů metody AE-akademická pracoviště Brno)**

| **Jméno přednášejícího** | **Počet** |
|---|---|
| Břetislav SKRBEK | 32 |
| Bernard KOPEC | 30 |
| *Jaroslav OBRAZ +* | 26 |
| Zdeněk PŘEVOROVSKÝ | 22 |
| Lubomír SODOMKA | 22 |
| Jaroslav FIALA | 21 |
| Miroslav PROCHÁZKA | 21 |
| *Oldřich TARABA +* | 18 |
| Václav SVOBODA | 18 |
| Karel HÁJEK | 18 |
| Zbyněk ZAVADIL | 18 |
| *Richard REGAZZO +* | 16 |
| Josef ŠIKULA | 14 |
| Václav BÍZEK + | 13 |
| Stanislav ŠTARMAN | 13 |
| Jiří BLAHŮŠEK | 12 |
| Jiří ŠNAJBERG | 11 |
| Jiří ULLMAN | 11 |
| Jan CRHA | 11 |
| Zdeněk SKÁLA | 11 |

**Tab. 2 Přehled autorů a počtu referátů na konferencích DEFEKTOSKOPIE 1997 až 2019 z metody AE-akademická pracoviště Brno**

| Jméno přednášejícího | Počet |
|---|---|
| Luboš PAZDERA | 59 |
| Pavel MAZAL | 45 |
| Marta Kořenská | 44 |
| Libor TOPOLÁŘ | 26 |
| Jaroslav SMUTNÝ | 18 |
| Ivan KUSÁK | 17 |
| František VLAŠIC | 17 |
| Michal MATYSÍK | 14 |
| Monika MATYSOVÁ | 12 |
| Miroslav LUŇÁK | 18 |

Poznámka: U jednoho referátu je uvedeno 4 až 8 autorů

**Tab. 3 Přehled míst konání DEFEKTOSKOPIE 1970 až 2019**

| ČECHY | | MORAVA | | SLEZSKO | |
|---|---|---|---|---|---|
| Praha | 17x | Znojmo | 3x | Ostrava | 3x |
| Tábor | 6x | Brno | 2x | Krásná | |
| Hradec Králové | 3x | Vsetín | | | |
| Mariánské Lázně | 3x | Zlín | | | |
| Karlovy Vary | | Olomouc | | | |
| Žďár nad Sázavou | | Luhačovice | | | |
| Liberec | | | | | |
| Špindlerův Mlýn | | | | | |
| Plzeň | | | | | |
| Seč u Chrudimi | | | | | |
| Chomutov | | | | | |
| České Budějovice | | | | | |

**Tab.4 Přehled zemí přednášejících účastníků DEFEKTOSKOPIE 1970-2019**

| Evropa | Asie | Afrika | Amerika |
|---|---|---|---|
| Československo, Česko, | Malajasie | Egypt | USA |
| Slovensko | Tchaiwan | Alžír | Kanada |
| Sovětský svaz,Rusko | Irán | | Brazílie |
| Estonsko | Korea | | |
| Ukrajina | Indie | | |
| Bělorusko | Turecko | | |
| Německá spolková republika, Německo | Japonsko | | |
| Polsko | | | |
| Maďarsko | | | |
| Bulharsko | | | |
| Rumunsko | | | |
| Slovinsko | | | |
| Belgie | | | |
| Francie | | | |
| Itálie | | | |
| Španělsko | | | |
| Řecko | | | |
| Švýcarsko | | | |
| Švédsko | | | |

# Využití dronů při diagnostických průzkumech stavebních objektů

Michal JANKŮ [1], Jiří GROŠEK [1], Josef STRYK [1]

[1] Centrum dopravního výzkumu, v. v. i., Brno, Česká republika

Phone: +420 773 644 321; e-mail: michal.janku@cdv.cz

**Abstrakt**
Bezpilotní letadla (někdy též UAV z anglického Unmanned Aerial Vehicle nebo také dron z anglického drone) byla ještě nedávno mnohými z nás považována za hračku pro malou skupinku nadšenců, případně se vědělo o jejich využití pro vojenské účely. Oblast dronů však v posledních letech zažívá dynamický rozvoj, co se týče nejen samotné techniky, ale i nových aplikací, ve kterých jsou drony používány. Jednou z nich je i diagnostický průzkum staveb. Drony umožnují snadno a relativně bezpečně prozkoumat stavební objekty v situacích, v nichž bylo ještě nedávno nezbytné riskovat nasazení pracovníků na výškové práce. Další výhodou dronů je možnost snadného získání snímků z ptačí perspektivy, což bylo dříve možné jen s použitím letadel či vrtulníků.

**Klíčová slova:** dron, bezpilotní letadlo, fotogrammetrie, diagnostický průzkum, Úřad pro civilní letectví (ÚCL)

**Abstract**
Not long ago, Unmanned Aerial Vehicles (UAVs sometimes called drones) were considered by many of us to be a toy for a small group of enthusiasts, or were known to be used for military purposes. However, the field of drones has experienced dynamic development in recent years, not only in terms of the technology itself, but also in terms of new applications in which drones are used. One of these is the diagnostic survey of buildings. Drones make it possible to survey construction sites easily and relatively safely in situations where, until recently, it was necessary to risk deploying personnel for high-rise work. Another advantage of drones is the ability to easily get a bird's eye view, which was previously only possible using aircraft or helicopters.

**Keywords:** drone, Unmanned Aerial Vehicles (UAV), photogrammetry, diagnostic survey, Civil Aviation Authority

## 1. Úvod

Díky rychlému vývoji v oblasti dronů, lze nyní pořídit za dostupnou cenu techniku, která byla ještě před několika lety v dané kategorii nemyslitelná. Za nižší desítky tisíc je v současné době možné koupit drony, se kterými lze natáčet profesionální záběry nebo pořizovat fotografie ve vysokém rozlišení. Zároveň jsou tyto drony vybaveny navigačními senzory a senzory detekce překážek, které usnadňují ovládání dronů a pomáhají je chránit před havárií. Základy pilotování se lze naučit během několika minut. Díky tomu se drony stávají dostupnější pro širší okruh uživatelů a rozšiřují se také možnosti jejich využití. V posledních letech se proto stává stále běžnější i jejich použití ve stavebnictví. Jako příklad lze uvést vizuální kontrolu obvodového pláště budov nebo plochých střech. Pokud se na dron připevní termokamera, lze s její pomocí lokalizovat místa, kde dochází k únikům tepla nebo pronikání vlhkosti do konstrukce. Pomocí dronů je také možné téměř automaticky pořizovat velké množství snímků, ze kterých lze s využitím fotogrammetrie vytvářet 3D modely budov, ortofoto mapy nebo pomáhat s určením objemu zemních prací. Na staveništi je možné prostřednictvím dronů monitorovat postup prací. Speciální drony dokážou létat i uvnitř budov, kde mohou pomáhat s diagnostickým průzkumem.

## 2. Létání s drony – legislativa

Se zvyšujícím se počtem provozovaných dronů však narůstá i potřeba jejich regulace. Od 31. prosince 2020 začala platit nová evropská pravidla regulující provoz bezpilotních letadel. Nově byly zavedeny 3 kategorie provozu: otevřená, specifická a certifikovaná. Asi největší změnou oproti předchozímu stavu je povinnost registrace. Neregistruje se dron, ale jeho provozovatel (zpravidla jím je vlastník dronu). Povinnost registrace se týká všech provozovatelů bezpilotních systémů, kteří splňují některou z následujících podmínek:

- provozují dron v „otevřené" kategorii s maximální vzletovou hmotností 250 g a vyšší nebo je tento dron vybaven čidlem schopným zachycovat osobní údaje (digitální kamera),
- provozují dron jakékoliv hmotnosti ve „specifické" kategorii.

Po provedení registrace obdrží provozovatel e-mailem dokument s přiděleným registračním číslem (poznávací značkou), kterým je povinen označit všechna provozovaná bezpilotní letadla. Dále byla zavedena povinnost registrace pilota bezpilotního systému včetně online testu z teoretických znalostí, jehož úspěšné absolvování je podmínkou pro provoz UAS v podkategoriích A1 a A3 otevřené kategorie provozu bezpilotních letadel. Pro provoz dronů v podkategorii A2 je nutné úspěšně absolvovat test na Úřadu pro civilního letectví (ÚCL) a stát se držitelem osvědčení o způsobilosti dálkově řídícího pilota.

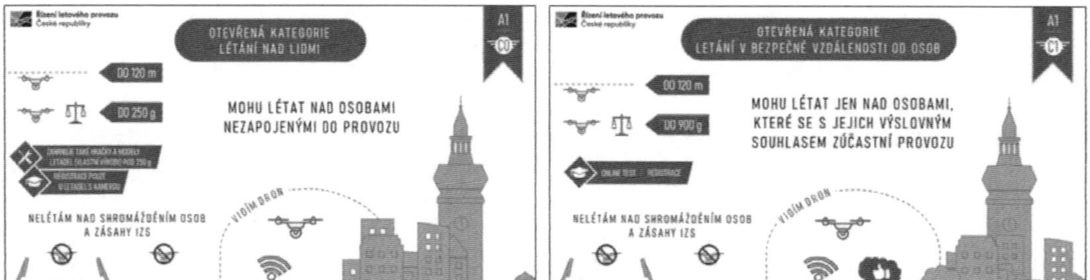

Obrázek 1. Pravidla pro létání s drony v Otevřené kategorii [1]

Jak je vidět z přehledu na obr. 1, nad urbanistickými oblastmi lze v otevřené kategorii létat pouze s dronem spadajícím do podkategorie A1, tedy lehčím než 250 g, jinak je nutné vlastnit osvědčení pro provoz v podkategorii A2. Podkategorie A2 je zaměřena na provoz, při kterém se předpokládá, že významná část letu bude prováděna v blízkosti lidí. Minimální vzdálenost od osob nezapojených do provozu se pohybuje od 5 m do 50 m. Vzdálenost 5 m je přípustná pouze v případě, že bezpilotní letadlo má zapnutou funkci nízko-rychlostního režimu. V ostatních situacích, ve kterých provozovatel dronu potřebuje překročit limity otevřené kategorie, spadá automaticky do specifické kategorie. Pilotovat dron v této kategorii mohou pouze držitelé příslušného Oprávnění k provozu vydávaného ÚCL. Dále také platí, že pokud se chce pilot pohybovat s dronem v ochranných pásmech dopravních staveb, nadzemních inženýrských sítí, vodních zdrojů, objektů důležitých pro ochranu státu a uvnitř zvláště chráněných území, potřebuje k tomu další oprávnění (vydané pro konkrétní činnost) potvrzené ÚCL s předchozím souhlasem příslušného správního orgánu či oprávněné osoby.

Výše zmíněná pravidla platí pro létání s drony ve vzdušném prostoru České republiky (Evropské unie), při provádění diagnostického průzkumu pomocí dronů uvnitř budov proto není potřeba na tato pravidla brát zřetel.

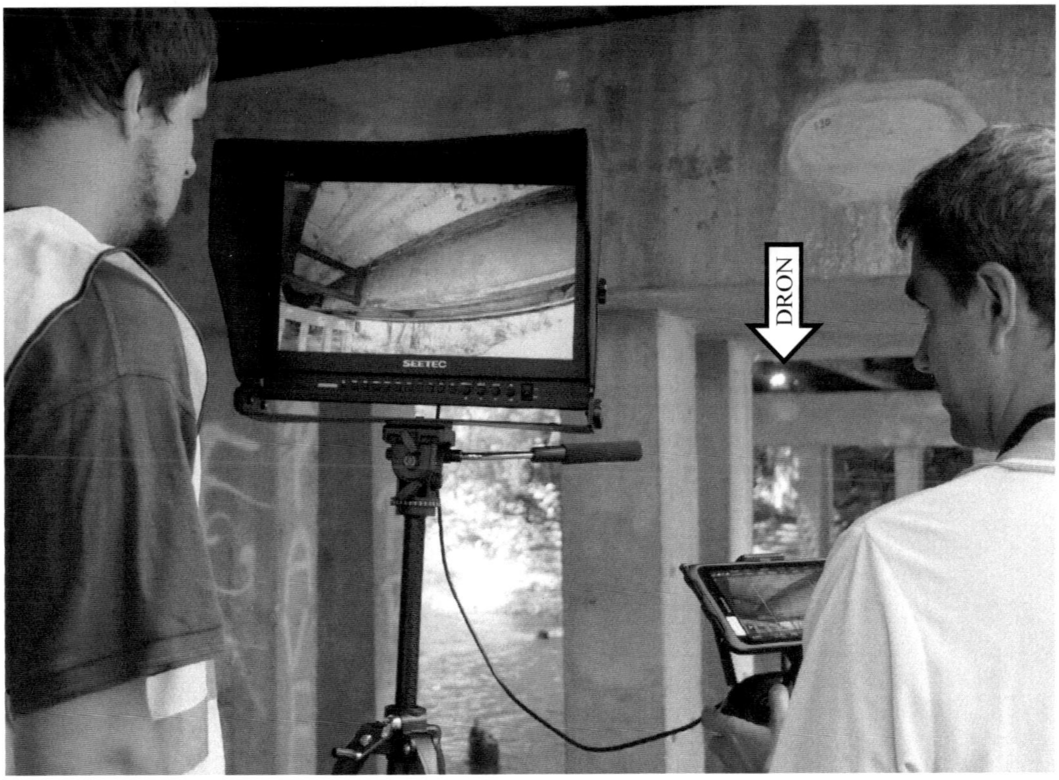


Obrázek 2. Diagnostický průzkum mostu pomocí dronu Flyability Elios 2

## 3. Provádění diagnostického průzkumu

Drony lze využít k průzkumu mnoha různorodých stavebních objektů jako jsou mosty, historické památky, továrny, haly, elektrárny atd. Například u mostů je kontrola stavebního stavu prováděna v rámci tzv. hlavních prohlídek, u kterých musí být zajištěno takové zpřístupnění konstrukce mostu, aby bylo možné podrobně zkontrolovat všechny jeho části a prvky přístupné bez zásahů do konstrukce. Nejvhodnější způsobem zpřístupnění při podrobné kontrole stavu mostní konstrukce je zpřístupnění tzv. „na dosah ruky".

Při prohlídkách hůře dostupných konstrukcí bývalo nutné využívat vysokozdvižnou plošinu či mostní prohlížečku. Pod každý most ovšem nejde jednoduše přijet s vysokozdvižnou plošinou a použití mostní prohlížečky komplikuje jejich omezená dostupnost v ČR, se kterou se pojí i vysoká cena za pronájem. Tento problém je možné vyřešit s pomocí dronů (obr. 2). Díky jejich použití lze zvýšit bezpečnost a efektivitu práce při provádění diagnostického průzkumu. Zároveň lze tímto způsobem provést kontrolu mostních objektů úsporněji a rychleji ve srovnání s použitím mostní prohlížečky [2]. Existují dokonce snahy tyto kontroly provádět pomocí automaticky řízených dronů, které mohou létat kolem konstrukcí (a pořizovat jejich fotografie) po předem naplánovaných trasách. Detekce poškození konstrukce na jednotlivých snímcích je provedena s využitím metod umělé inteligence pomocí softwaru založeném na tzv. neuronové síti [3,4]. Nicméně tyto projekty jsou zatím řešeny spíše pro vysoké mosty, kolem jejichž pilířů mají drony dostatek manévrovacího prostoru. V ostatních případech je potřeba drona řídit manuálně a létat v přiměřené vzdálenosti, tak aby nedošlo ke kontaktu vrtule dronu s konstrukcí mostu nebo s jiným objektem v jeho blízkosti. Větší vzdálenost od objektů lze kompenzovat osazenou kamerou s dostatečným přiblížením. Například dron DJI Matrice 300 lze doplnit senzorem, který v sobě zahrnuje kameru s 23x optickým zoomem, širokoúhlou kamerou, laserovým měřičem vzdálenosti a termokamerou (obr. 3).

Obrázek 3. Dron DJI Matrice 300

Obrázek 4. Dron Flyability Elios 2

Pokud je potřeba pořídit snímky z větší blízkosti, je možné snížit riziko poškození dronu tím, že se na něj dodatečně připevní kryty vrtulí. Některé speciální drony jsou už z výroby navrženy pro let ve vnitřních prostorech, jako například Elios 2 od firmy Flyability (obr. 4). Jelikož je navržen pro průzkum vnitřních prostor, nepotřebuje tento dron GNSS senzor, namísto toho je vybaven IMU jednotkou, barometrem, magnetometrem a sedmi dalšími senzory, které mu pomáhají snadno se pohybovat v prostoru nebo naopak „viset" na místě. Dron je vybaven výkonným přisvícením, díky čemuž může létat v různých tmavých prostorech, jako jsou například potrubí, komíny, skladovací nádrže, důlní štoly atd. [5]. Ochranná klec z uhlíkových vláken je dimenzována na náraz do pevné překážky až při rychlosti 3 m/s bez poškození dronu.

V případě historických objektů může existovat snaha o co nejvěrnější zaznamenání jejich aktuálního stavu. I v tomto případě mohou pomoci drony, se kterými lze automatizovaně pořizovat velké množství fotografií nejen exteriérů, ale též interiérů různých památek. Aby dron věděl, kde má létat, je potřeba nejprve vnitřní prostory naskenovat pomocí pozemního laser skenu [6]. V počítači se pak vytvoří 3D model interiéru a naplánuje se trajektorie letu dronu tak, aby vyfotografoval všechna požadovaná místa. Během letu dron využívá nejen uloženou 3D mapu okolního prostředí, ale též senzory, které mu pomáhají při orientaci v prostoru. Takto lze například podrobně zdokumentovat vnitřní prostory kostelů mnohem rychleji, než kdyby se mělo stavět lešení a fotografie vnitřních stěn pořizovat z něj [7]. Ze získaných snímků je možné následně pomocí fotogrammetrie vytvářet například texturované 3D modely. Podobně se dají zpracovat fotografie (exteriéru) téměř jakéhokoliv objektu nasnímaného i z běžně dostupného dronu (obr. 5).

Obrázek 5. Texturovaný 3D model kostela Panny Marie Pomocné na Uhlířském vrchu nedaleko Bruntálu

## 4. Závěr

Autoři článku využívají drony především při prohlídkách a kontrole stavebního stavu mostů. V článku bylo nastíněno i několik dalších způsobů jejich použití při diagnostickém průzkumu ostatních stavebních objektů, možnosti dronů jsou však daleko větší. Autoři článku předpokládají, že jejich význam bude v budoucnu dále narůstat, například při přepravě zboží a osob. Každá mince má však dvě strany, proto lze jen doufat, že budoucí pokrok v tomto odvětví přinese více pozitiv než negativ.

**Poděkování**

Tento článek je financován se státní podporou Technologické agentury ČR a Ministerstva dopravy v rámci Programu DOPRAVA 2020+, v rámci řešení projektu CK01000108 Nové přístupy při diagnostice stavu výztuže předpjatých mostních nosníků.

**Literatura**

1. Otevřená kategorie. Létejte zodpovědně [online]. Řízení letového provozu České republiky [cit. 2021-8-26]. Dostupné z: https://letejtezodpovedne.cz/legislativa/jak_letame_nyni/kategorizace_provozu?clid=150

2. Diagnostický průzkum a prohlídky mostů. Centrum dopravního výzkumu [online]. Brno, 2021 [cit. 2021-8-26]. Dostupné z: https://www.cdv.cz/diagnostika-a-prohlidky-mostu/

3. M Friese and R Holst, 'Inspection of engineering structures and digital technologies to detect and analyse damages', Annual Report 2019: Reports of the Federal Highway Research Institute [online]. Bergisch Gladbach: Bundesanstalt für Straßenwesen, 2020, pp 26-27. ISBN 978-3-95606-499-9. Dostupné z: https://www.bast.de/BASt_2017/DE/Publikationen/Berichte/unterreihe-a/2020-2018/A42.html?nn=1836242

4. P Furtner, E Forstner, A Karlusch, M Luňáček and Z Jeřábek, 'Automatizované inspekce dopravní infrastruktury na bázi digitálních dvojčat a strojového učení', Sanace 2019, Fakulta stavební VUT v Brně, 23.–24.5. 2019, pp 67-76, ISBN 978-80-214-5749-2.

5. Articles & Media. Flyability [online]. Paudex, Switzerland [cit. 2021-8-29]. Dostupné z: https://www.flyability.com/articles-and-media

6. Specializovaná platforma. Dron & Monument [online]. Praha, 2021 [cit. 2021-8-29]. Dostupné z: https://dronument.cz/platforma

7. P Petráček, V Krátký and M Saska, 'Dronument: System for Reliable Deployment of Micro Aerial Vehicles in Dark Areas of Large Historical Monuments', IEEE Robotics and Automation Letters, Vol 5, No 2, pp 2078-2085, January 2020.

# Možnosti nedestruktivního zkoušení křemíkových materiálů

Pavel MAREŠ [1], Jana VESELÁ [1]

[1] odd. Diagnostika a kvalifikace, Centrum výzkumu Řež
Husinec - Řež, Česká republika; Tel.: +420 606 138 199, e-mail: pavel.mares@cvrez.cz, jana.vesela@cvrez.cz

**Abstrakt**
Nedílnou součástí počítačů jsou procesory, skládající se z tranzistorů, které jsou vyrobeny pomocí složitých metod z křemíku, které vyžaduje znalost vnitřní kvality materiálu. V oblasti nedestruktivního zkoušení bylo testováno několik metod, které by mohly být případně vřazeny do výrobního postupu tak, aby vnitřní vady byly detekovány v rané fázi výroby. Mezi tyto metody patří zejména ultrazvukové zkoušení s využitím imerzní techniky. V rámci vývojových prací bylo provedeno několik testování křemíkových ingotech od rentgenového zkoušení až po automatizované ultrazvukové zkoušení v imerzním tanku.

**Abstract**
An integral part of computers are processors, consisting of transistors, which are made using complex methods of silicon, which requires knowledge of the internal quality of the material. In the field of non-destructive testing, several methods were tested, which could possibly be incorporated into the production process so that internal defects are detected at an early stage of production. These methods include, in particular, ultrasonic testing using immersion techniques. As part of the development work, several tests of silicon ingots were performed, from X-ray testing to automated testing in an immersion tank.

**Klíčová slova:** křemík, ultrazvuk, imerzní zkoušení, výrobní vady, automatizované zkoušení

## 1. Úvod

Testování křemíkových ingotů bezprostředně po výrobě může vést k odhalení nežádoucích vad uvnitř materiálu a tím předejít jejich pozdnímu odhalení až při dalších fází výroby. Proto bylo provedeno testování vhodných metod nedestruktivního zkoušení, které mají za cíl detekci možných vad vzniklých výrobě. Jako další kritérium pro výběr vhodné metody je i její možnost začlenění přímo do výroby a snížení časové náročnosti na takovou kontrolu.

## 2. Základy technologie výroby

Nejprve se provede chemická reakce křemenného písku s uhlíkem, kdy vznikne tzv. hutní křemík a oxid uhelnatý (dle reakce $SiO_2 + 2C \rightarrow Si + 2CO$). Hutní křemík však není dostatečně čistý, provede se proto reakce s chlorovodíkem a vznikne trichloristan ($Si + 3HCl \rightarrow SiHCl_2 + H_2$). Ten se destilací opětovně čistí a po další reakci s vodíkem vzniká dostatečně čistý křemík vhodný pro použití v integrovaných obvodech. Vyčištěný křemík má polykrystalickou strukturu (materiál je složený z malých pravidelně uspořádaných zrn s různou vzájemnou orientací), která je pro použití v integrovaných obvodech nevhodná.
Pro elektrotechniku je vyžadován monokrystal, který má shodnou orientaci krystalografické mřížky v celém svém objemu. Monokrystal se vyrábí přetavením polykrystalu s následnou přesně řízenou krystalizací pomocí Czochralského metody v zařízení nazvaném tažička. Nejprve se střepy polykrystalického křemíku nasypou do kelímku (ve skutečnosti jsou dva — vnitřní a vnější). Přidáno může

být také požadované množství příměsi, často se totiž nevyrábí dokonale čisté monokrystaly, ale monokrystaly s nevlastní vodivostí typu N nebo P.
Vnitřní kelímek je vyroben z křemenného skla a přímo v něm je uložen polykrystalický křemík. Křemenné sklo je zvoleno proto, že s roztaveným křemíkem reaguje jen velmi pomalu a současně je schopné odolat teplotám okolo 1500 °C. Tento kelímek je pak vložen do vnějšího grafitového kelímku, který je zahříván taktéž grafitovým topným tělesem a rovnoměrně rozvádí teplo do vnitřního kelímku.
Teplota taveniny se udržuje jen lehce nad teplotou tání křemíku okolo 1450 °C.
Jakmile je křemík roztavený, pomalu se do něj zanoří držák se zárodkem. Zárodek je malý monokrystal se stejnou krystalografickou orientací, jakou požadujeme u výsledného monokrystalu. Ponořená část zárodku se v tavenině začne taktéž tavit, čímž dojde k roztavení případných vad způsobených předchozím použitím zárodku a současně se vytvoří vazba na taveninu (semínko — seed). Zárodek se semínkem se začíná vytahovat z taveniny, přičemž rychlost vytahování je výrazně vyšší než při samotném růstu monokrystalu. Vzniká tak úzký krček s průměrem jen několik milimetrů a délkou několik centimetrů. Jeho účelem je odstranit případné poruchy krystalografické mřížky (dislokace), které mohly vzniknout teplotním šokem při prvotním ponoření zárodku do taveniny. Dislokace se pod velkým úhlem šíří od středu materiálu k jeho povrchu, kde zanikají (proto je krček tak úzký)
Jakmile je monokrystal zarovnán a je nastaven požadovaný průměr krystalu (v dnešní době typicky 300 mm), probíhá tažení samotného těla monokrystalu, ze kterého se následně vyrábí integrované obvody. Tažení musí probíhat v co nejvíce homogenních (stejnoměrných) podmínkách a sledují se parametry jako je teplota taveniny (cca 1450 °C v objemu, na přechodu tavenina/monokrystal se udržuje teplota tuhnutí křemíku — 1413 °C), rychlost tažení monokrystalu (jednotky milimetrů za hodinu), rychlost otáčení kelímku a monokrystalu (protiběžné, používá se kvůli zachování co možná nejvyšší homogenity taveniny) a tlak a koncentrace ochranné plynné atmosféry (většinou argonové).
Tělo monokrystalu se táhne po dobu několika desítek hodin a je dlouhé okolo jednoho metru.
Po vytažení těla monokrystalu následuje tvorba špice krystalu, kdy se rychlost tažení postupně zvyšuje, čímž klesá průměr monokrystalu. Zúžení zamezuje vzniku poruch krystalové mříže, ke kterému může dojít vlivem teplotního šoku po vytažení celého krystalu z taveniny. Křemíkový monokrystal se proto následně chladí jen pomalu.
Hotový křemíkový monokrystal se nazývá ingot obr. 1, odstřihne se z tažičky u krčku a je připraven k dalšímu zpracování. Výroba ingotu z polykrystalického křemíku většinou trvá dva až tři dny, ingot může být delší než metr a půl s hmotností pře 100 kg.

Obrázek 1. Hotové monokrystaly různých průměrů (*dreamstime.com*)

## 3. Nedestruktivní zkoušení

Snaha získat informace o vnitřní kvalitě taženého ingotu v průběhu výrobní etapy dělení na dílčí části těla ingotu vedlo k ověřování objemových metod NDT jakými jsou ultrazvukové zkoušení (UT), rentgenová zkouška (RT) v kombinaci s povrchovými metodami jakou jsou vizuální kontrola (VT), zkoušení vířivými proudy nebo infračervená termografie (IR).

Nejprve byl ke zkouškám dodán vzorek o vnějším průměru cca 218 mm, tloušťce 19,5 mm. Čelní strany byly označeny jako A a B a byl vyznačen střed vzorku křížkem. Na tomto vzorku bylo provedeno testování všemi výše zmíněnými metodami.
Následně bylo provedeno ultrazvukové zkoušení imerzní technikou na křemíkových válcích tak, jak by mělo probíhat přímo ve výrobním závodě. Válce měly jak opracovaný povrch, tak neopracovaný a jejich průměr byl 150 resp. 160 mm a 200 resp. 210 mm.

### *2.2 Vizuální zkoušení*

Touto metodou byly zjištěny dvě vady označené čísly 1 a 2. Vada č. 1 na straně A, vada č. 2 na straně B, jejich vzdálenosti od vnějšího průměru jsou uvedeny na obr. 2.

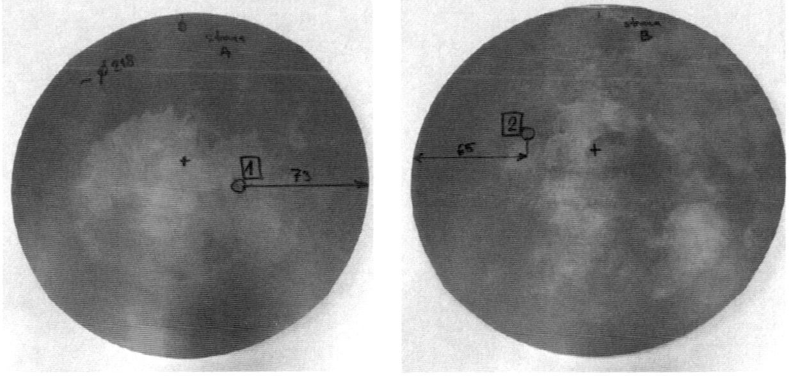


Obrázek 2. Vzorek strana A (vlevo) a strana B (vpravo)

### *2.2.1 Ultrazvukové zkoušení*
### *2.2.1 Vzorek typu disk*

Bylo provedeno konvenční ultrazvukové zkoušení technikou PEUT (pulse-echo). S ohledem na dobrou prozvučitelnost vzorku Si ingotu byly ověřovány 2 přímé ultrazvukové sondy s frekvencí 4 a 10 MHz. Při zkoušení z obou čelních povrchů byla detekována indikace vnitřní vady, označena jako č. 3. Její odezva na A-scanu je zobrazena na obrázku 3.

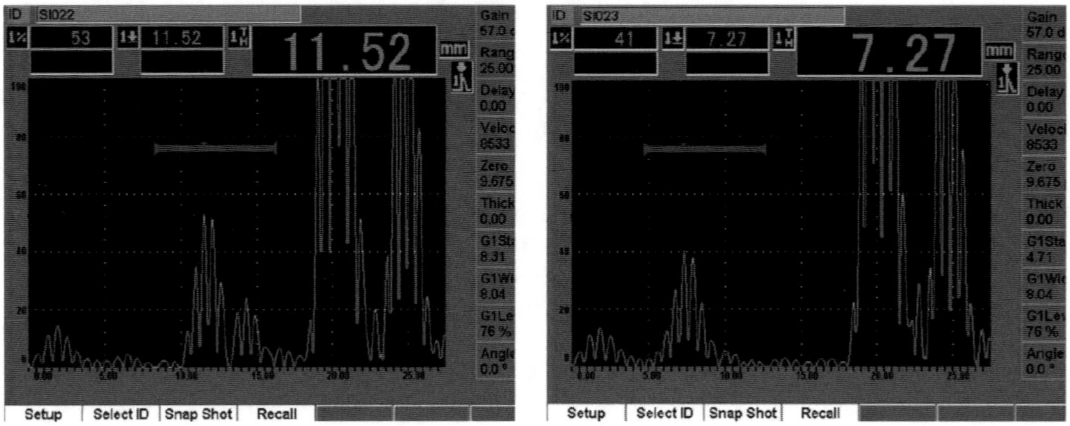


Obrázek 3. Vzorek strana A (vlevo) a strana B (vpravo)

### *2.2.2 Vzorky křemíkových válců*

Pro ověření možné přítomnosti defektů ve vzorcích bylo aplikováno mechanizované imerzní ultrazvukové zkoušení a radiografické zkoušení. Specifikem zkoušení Si válců je rychlost šíření podélných vln v materiálu, která je ve srovnání s běžnými materiály v průmyslu vyšší a dosahuje hodnot v rozsahu 8500 až 9000 m/s.

Během imerzního zkoušení jsou sondy umístěny ve vodě nad, nebo vedle zkoušeného dílu a ultrazvukový svazek sondy se promítá do zkoušeného kusu. Celé zkoušení probíhá v imerzní nádrži s vodou. Ovládání umožňuje operátorovi nastavit rychlost otáčení stolu (0-30 otáček za minutu), náklon sondy, a pak automaticky sledovat (krokovat) rovinu vodorovnou, svislou či šikmou. Používají se speciální imerzní sondy, nejlépe v rozsahu frekvence 2,5 - 10 MHz s průměry měničů 0,375" - 0,75" a fokusací do hloubky 3"-6".

V našem případě by cílem zkoušení bylo ověřit vhodnost zkušební techniky pro detekci vnitřních vad.

Obrázek 4. Umístění válce ve zkušební nádrži

Na provedení UT má zásadní vliv prozvučitelnost materiálu, která je v případě Si ingotů velmi dobrá a rychlost šíření ultrazvuku je zhruba 9000 m/s. Pro imerzní zkoušení je důležité mít rovnoběžná čela válce z důvodu zpřesnění polohy indikací a také možnosti vyhodnocovat pokles UT KE.
Na válcích č. 2 (6") a č. 3 (8") byl z čela v axiální ose každého z válců vyroben vývrt o průměru 5,47 mm do hloubky 100 mm. Vývrt s plochým jako náhradní vada je určený k nastavení citlivosti zkoušení.

Celkem bylo na opracovaném válci o průměru 6" detekováno 8 indikací a na válci s neopracovaným povrchem 11 významných indikací s ohledem na poměr amplituda / šum. Na obou 8" válcích č. 3 a 4 nebyly detekovány žádné indikace při zachování poměru šum/amplituda min 6 dB, a to bez ohledu na stav válcového povrchu.
Z uvedených výsledků ultrazvukového zkoušení imerzní technikou je zřejmé, že vnitřní vady / indikace byly detekovány na obou 6" válcích č. 1 a 2, a tedy bez vlivu stavu vnějšího válcového povrchu, tj. povrch válce opracovaný nebo neopracovaný nemá vliv na detekci indikací.

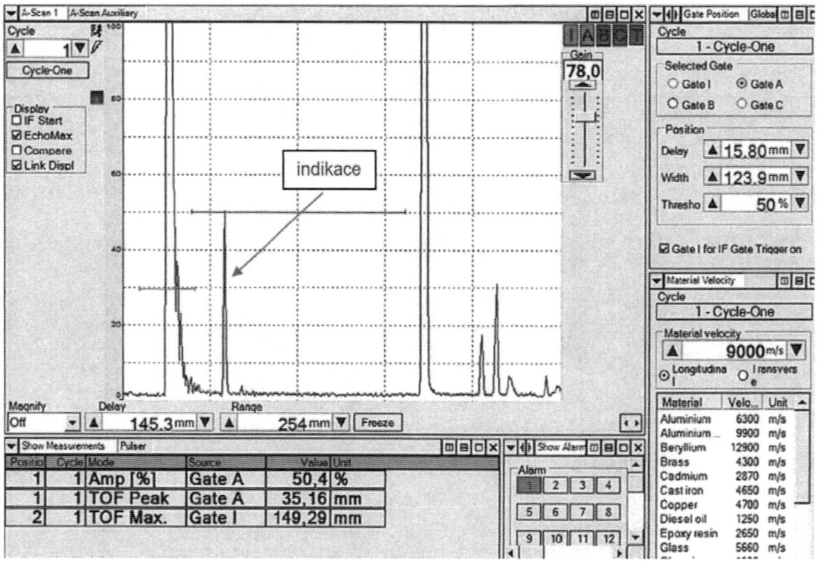


Obrázek 5. Příklad detekce indikace na válci 6" s opracovaným povrchem

Imerzní technika je zvláště vhodná pro zkoušení jednotlivých objektů, což se potvrdilo o v případě křemíkových monokrystalů ve formě válců. S vhodně zvolenou frekvencí sondy je možné detekovat vnitřní vady vznikající při výrobě. Z celkového počtu 4 válců byly indikace nalezeny na válcích č. 1 a 2, bez ohledu na stav povrchu.

### *2.3 Rentgenové zkoušení*

Při testování rentgenového zkoušení byl použit zdroj RSCO MF4(rentgenka) od firmy GE Inspection Technologies s výkonem max 900W a zkušebním napětím 5–200 kV. Deskový detektor DX 250UW pro snímkování a SW Rythm review pro následné obrazové zpracování dat.
Při tomto zkoušení byly detekovány dvě povrchové vady a jedna objemová indikace, viz obrázek 6.

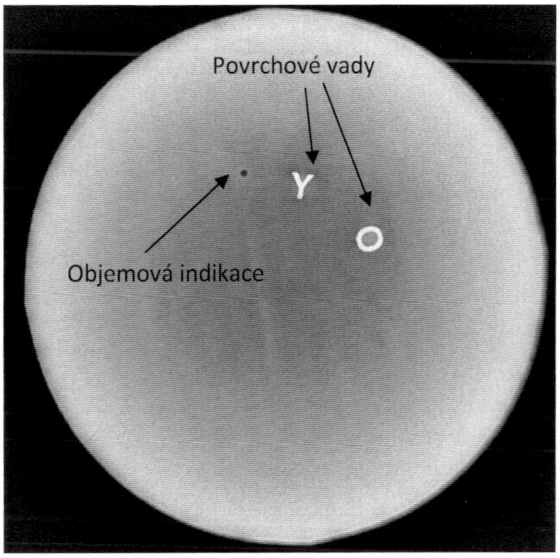


Obrázek 6. Příklad detekce indikace na válci 6" s opracovaným povrchem

### 2.4 Infračervená termografie

Pro zkoušení z obou povrchů byla použita flash kamera a parametry byly optimalizovány vzhledem k typu vzorku a kvalitě povrchu. Vady komunikující s povrchem jsou pomocí SW dobře detekovány a vůči pozadí rozlišeny. IR metoda zkoušení je schopna detekovat vady i pod povrchem cca do hl. 1-2 mm.

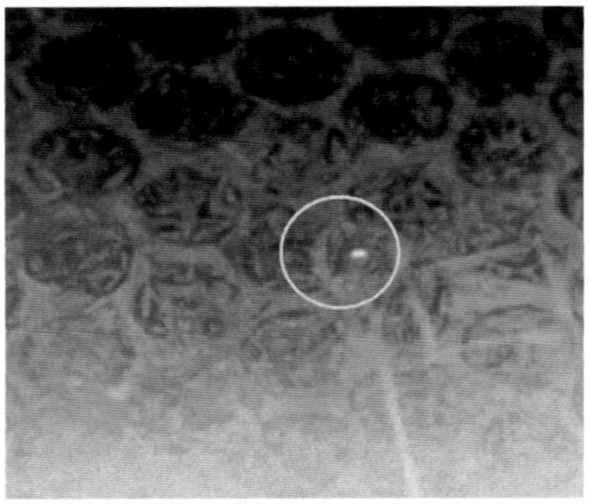

Obrázek 7. Příklad detekce indikace na válci 6" s opracovaným povrchem

### 2.5 Zkoušení vířivými proudy

Pro vyloučení možnosti využívat metodu ET ke zjištění povrchových a podpovrchových vad bylo provedeno měření pomocí flexibilní sondy s využitím techniky pole vířivých proudů. Při tomto měření bylo zjištěno, že ani jedna povrchová vada nebyla detekována.

## 3. Závěr

Povrchové vady typu póru lze detekovat metodami VT a IR. Pro detekci povrchových, případně podpovrchových vad nelze použít metodu ET z důvodu typu zkoušeného materiálu. Vady uvnitř materiálu lze detekovat zejména metodou UT. Materiál všech vzorků je velmi dobře průchodný pro ultrazvukové vlny, což umožňuje použití sond s vyšší frekvencí, a tedy i vyšší vyhledávací citlivostí. Přesnost detekce vady je závislá na kalibraci rychlosti šíření ultrazvukových vln.
Pro zařazení do výrobního procesu se jeví jako nejvhodnější využití imerzní techniky zkoušení s využitím zkušební vany a automatizovaného polohování ultrazvukové sondy.

**Poděkování**
Prezentované výsledky byly realizovány v rámci Institucionální podpory Ministerstva průmyslu a obchodu.

**References**

References should be written in the order in which they appear in the text in the following format:

1. T. Šulc, 'Od písku k procesoru – výroba křemíkového waferu', PCTuning, červen 2020 [1]
2. J. Veselá, P. Mareš, 'Si válce – výsledky nedestruktivního zkoušení', Technická zpráva, Centrum výzkumu Řež s.r.o., Plzeň, říjen 2020 [2]
3. J. Veselá, 'Ověření NDT metod pro zkoušení křemíkových ingotů', Technická zpráva, Centrum výzkumu Řež s.r.o., Plzeň, leden 2020 [3]

# Zvyšování a ověřování odborné způsobilosti pracovníků NDT provádějících zkoušky na vybraných zařízeních jaderných elektráren.

Josef NEUGEBAUER [1], Petr FOUSEK [2]

[1] TÜV NORD Czech, s.r.o.

e-mail: neugebauer@tuev-nord.cz

[2] ČEZ a.s. Divize jaderná energetika, Email: petr.fousek01@cez.cz

## Abstrakt

V lednu 2021 vyšla revize Normativně technické dokumentace A.S.I. sekce VII, týkající se NDT kontrol jaderných elektráren typu VVER. Již od začátku tato dokumentace obsahuje požadavek na ověřování znalostí personálu NDT formou atestací. Držitel povolení k provozování JE, firma ČEZ, se k tomuto úkolu postavila čelem a zahájila atestaci svých i externích pracovníků provádějící zkoušení na vybraných zařízeních JE. Tento článek podává přehled o tom, co tyto atestace obsahují, jak a kým se provádějí a jaká je jejich legislativní základna.

Klíčové slova: personál NDT, jaderná elektrárna, atestace personálu

## Abstract

A new revision of the A.S.I. Section VII, concerning NDT inspections of nuclear power plants was issued in January 2021. From the very beginning, this documentation contains a requirement to verify the knowledge of NDT personnel by attestations. ČEZ, the NPP operating license holder, began attesting their own and external employees performing testing on selected NPP equipment. This article provides an overview of what these attestations contain, how and by whom they are carried out, and what their legislative basis is.

Key words: NDT personnel, nuclear power plants, attestation of NDT personnel

## 1. Úvod

Zvyšování odborné úrovně zaměstnanců je v zájmu každé rozumné firmy. Existují různé možnosti jak zajistit odborný růst zaměstnanců. Pravděpodobně nejrozšířenější je vyslání pracovníků na některý z různých typů školení zaměřených na legislativní požadavky (např. školení BOZP) nebo různé kvalifikační kurzy podle požadavků standardů a norem (např. ISO 9712).
Tato školení mohou probíhat u externích organizací nebo si je může zaměstnavatel organizovat sám. Samozřejmým výstupem těchto školení je ověření nabitých znalostí. Toto ověření může být jen prosté přezkoušení formou jednoduchého testu bez kritérií, nebo komplexnější prověření i s kritérii jejichž splnění je nezbytné pro další výkon činnosti.
Společnost ČEZ a.s. Divize jaderná energetika (dále DJE) se jako Držitel povolení zabývá provozem jaderných elektráren na našem území. Z tohoto důvodu je vysoká odborná způsobilost pracovníků podílejících se na provozu a kontrolách jaderného zařízení nezbytná. Tento příspěvek se zabývá způsobem, jakým je ve společnosti DJE zajištěno zvyšování a ověřování odborné způsobilosti zaměstnanců podílejících se na provádění NDT.

## 2. Legislativní rámec

NDT je v rámci provozu a údržby jaderných zařízení jedním z důležitých bezpečnostních nástrojů. Obor NDT se neustále vyvíjí, vznikají nové techniky, postupy a dokonce i nové metody zkoušení. Tento neustálý vývoj, který významným způsobem vylepšuje nebo zpřesňuje výsledky kontrol a umožňuje získávat nové a přesnější informace o stavu kontrolovaných zařízení značným způsobem zvyšuje úroveň bezpečnosti zkoušených zařízení.

Jedním z klíčových dokumentů upravující činnosti při provádění a hodnocení nedestruktivních kontrol je dokument s trochu tajuplným názvem NTD A.S.I. sekce VII. Jde o Normativně Technickou Dokumentaci Asociace Strojních Inženýrů (NTD A.S.I.). Tato technická dokumentace se skládá z devíti sekcí. Každá sekce se zabývá konkrétními činnostmi prováděných na jaderných elektrárnách typu VVER.

Jak je zřejmé z předešlého textu, sekce VII se zabývá nedestruktivními kontrolami jaderných elektráren. Tento dokument specifikuje provádění a hodnocení (kritéria přípustnosti) při výrobě, montáži i provozu systémů, konstrukcí a komponent jaderných zařízení.

Další z dokumentů, které by měly být zmíněny je Vyhláška č. 358 z roku 2016 o požadavcích na zajišťování kvality a technické bezpečnosti a posouzení a prověřování shody vybraných zařízení. Kromětoho, že tato vyhláška obsahuje seznam vybraných zařízení, rozsah a způsob zajišťování kvality vybraných zařízení a technické požadavky na vybraná zařízení v Příloze 2, Kapitole 7, odstavec 7.2 uvádí následující:

*„Pracovníci provádějící nedestruktivní kontroly nerozebíratelných spojů musí být odsouhlaseni autorizovanou osobou, nejde-li o nerozebíratelný spoj na vybraném zařízení uvedeném v §12 odst. 3".*

Požadavek „odsouhlasení autorizovanou osobou" byl následně implementován právě do NTD A.S.I. sekce VII. A to konkrétně do Kapitoly 7, která se nazývá „Atestace personálu – požadavky na prokázání způsobilosti pro držitele povolení". V této kapitole se mimo jiné uvádí:

*„Atestační proces řídí a plně za něj zodpovídá Držitel povolení. Požadavky na atestaci musí být popsány v dokumentaci vydané Držitelem povolení".*

Pro splnění tohoto požadavku vypracoval Držitel povolení interní dokument s názvem „ Provádění atestace NDT personálu v DJE".

## 3. Zvyšování a ověřování odborné způsobilosti pracovníků NDT pomoc atestačních zkoušek.

### 3.1 Zvyšování odborné způsobilosti

Jak je uvedeno v předchozí kapitole, je požadavek na zvyšování a ověřování odborné způsobilosti zakotven v legislativě ČR.

Držitel povolení, v tomto případě ČEZ a.s. Divize jaderná energetika, tedy vypracoval interní metodiku, ve které implementoval výše uvedené požadavky.

Jak tedy takové zvyšování a ověřování odborné způsobilosti pracovníků NDT vypadá v podmínkách ČEZ. První otázka, která nás v tomto případě zajímá, zní, pro koho je tato atestace určena? Atestace se týká zaměstnanců ČEZ a dodavatelských organizací, kteří provádějí nedestruktivní zkoušení svarových spojů.

Podmínky přístupu k atestacím jsou tyto:

- platný certifikát dle ISO 9712 v dané metodě a v sektoru obsahující svary
- platná zraková způsobilost podle podmínek ISO 9712
- absolvování teoretické a praktické přípravy

První dva body požadavků, tedy platný certifikát a platná zraková způsobilost jsou jasné a notoricky známé. Co si však představit pod teoretickou a praktickou přípravou.

### 3.1.1 Teoretická příprava

Teoretická příprava, kterou uchazeč absolvuje je společná pro všechny NDT metody. Obecně se dá říci, že v rámci teoretické přípravy je uchazeč seznámen s veškerou dokumentací vztahující se k provádění kontrol. A se změnami v této dokumentaci zavedenými. Teoretická příprava probíhá formou dialogu lektora s uchazeči (na tomto místě je třeba podotknout, že počet uchazečů teoretického i praktického je striktně omezen na nejvýše 6 osob), kdy lektor probírá jednotlivé dokumenty a uchazeči se na ně obrací s připomínkami a dotazy vzešlé z jejich předešlé praxe.

I když takto popsaný proces přípravy u většiny lektorů vzbuzuje smíšené pocity vycházející s jejich dosavadní praxe, kde o nějakém dialogu většinou nemůže být řeč, v tomto případě to však funguje. Uchazeči o atestaci jsou dostatečně motivováni, aby projevili aktivitu a jejich připomínky, pokud jsou skutečně významné, jsou obratem zavedeny do příslušné dokumentace. Tato oboustranná výměna, ano dalo by se říci, že i synergie, je výhodná pro všechny zúčastněné.

Kromě debaty nad nutnou a požadovanou dokumentací je u uchazečů v rámci teoretické přípravy neméně důležité i prohloubení znalostí hodnotících výrobkových norem.

### 3.1.2 Praktická příprava

V rámci praktické přípravy uchazeči provádějí zkoušení na vybraných školících vzorcích, které obsahují reprezentativní vady, vyskytující se v provozu jaderných elektráren. Samozřejmostí je vždy vyhodnocení zkoušky dle zadaných kritérií i zápis této zkoušky do interní aplikace. Zde opět probíhá výměna zkušeností, kdy lektor především sjednocuje různé „krajové“ zvláštnosti při vypracování protokolů. Tímto vlastně dochází k uniformizaci zápisu výsledků zkoušek v rámci celé Divize Jaderné Energetiky.

## 4 Ověřování odborné způsobilosti

Ověřování odborné způsobilosti je prováděno atestační zkouškou.

Atestační zkouška se skládá ze dvou částí:

- teoretická část
- praktická část

Teoretická část zkoušky je prováděna formou testu obsahující otázky vybrané z aktuální a pravidelně doplňované banky testových otázek.

Tato zkouška je zaměřena na znalosti dokumentů, kterými se provádění NDT řídí. Je společná pro všechny metody a její platnost je dvakráte tak dlouhá jako platnost zkoušky praktické.

Praktická část zkoušky zahrnuje provedení kontroly na slepých vzorcích, jež jsou vázány na konkrétní metodu nebo inspekční postup. Součástí praktické zkoušky je provedení výstupního protokolu a zadokumentování výsledku do interní aplikace ČEZ.

Pro posouzení teoretických znalostí a praktických dovedností je pro každou atestační zkoušku schválena Atestační komise.

Atestační komise je nezávislá odborná komise. Členy atestační komise jsou:

- **Předseda atestační komise** – zaměstnanec certifikovaný minimálně ve stupni 3 v metodě, ve které je prováděna atestace a sektoru MS dle EN ISO 9712 pro praktickou část zkoušky, který odpovídá za provedení atestační zkoušky.
- **Dohled u atestační zkoušky** – zaměstnanec certifikovaný minimálně ve stupni 2 v metodě, ve které je prováděna atestace a sektoru MS dle EN ISO 9712 v případě potřeby pověřený předsedou atestační komise pro dohled nad zkouškou.
- **Autorizovaná osoba** (inspektor Autorizované osoby) – provádí odsouhlasení NDT personálu v DJE provádějících atestační zkoušky dle vyhlášky č. 358/2016 Sb., příloha 2, odstavec 7.2.
- **Nezávislá osoba** – účastní se atestační komise na základě rozhodnutí Garanta zvláštních procesů.

## 4.1 Hodnocení znalostí uchazečů

Teoretická i praktická část zkoušky se hodnotí samostatně

### 4.1.1 Teoretická část

Hodnocení testu se provádí standardně jako podle ISO 9712. To znamená, že správná odpověď odpovídá jednomu bodu. Uchazeč úspěšně absolvoval tuto část zkoušky, pokud jeho výsledek v procentech je rovný nebo lepší jako 80%.

Pokud uchazeč není úspěšný, nepokračuje dále k praktické části a může po uplynutí 14 dnů zkoušku zopakovat. Povolené jsou dvě opakování. Poté je na uchazeče nahlíženo jako na nového a smí si zkoušku zopakovat až po 12 měsících a absolvování obou částí přípravy.

Úspěšný uchazeč postupuje dále k vykonání praktické části zkoušky.

### 4.1.2. Praktická část

Hodnocení praktické části zkoušky provádí Předseda atestační komise za účasti inspektora Autorizované osoby. Hodnocení je prováděno porovnáním výsledného protokol vystaveného uchazečem se vzorovým protokolem konkrétního vzorku.

Hodnocení je prováděno pomocí procentuálních váhových faktorů, kdy největší důraz je kladen na zjištění vad, jejich záznam (sběr dat) a vyhodnocení. Proto, aby uchazeč splnil požadavky praktické části zkoušky, musí dosáhnout minimálně80%.

## 4.2 Celkové hodnocení

Celkové hodnocení obou částí atestační zkoušky se zaznamenává do Souhrnného protokolu, který schvaluje předseda atestační komise spolu s inspektorem Autorizované osoby, popřípadě ještě nezávislá osoba (pokud se atestací zúčastnila).

# 5 Atestace

Na základěsplnění požadavků na atestaci pracovníka. To znamená absolvování teoretické i praktické části přípravy a úspěšné složení obou částí zkoušky vystaví Autorizovaná osoba uchazeči *„Osvědčení NDT personálu v DJE*". Platnost tohoto Osvědčení je 18 měsíců. Poté musí uchazeč absolvovat celý proces znovu.

Tímto administrativním úkonem je ukončen proces zvyšování a ověřování odborné způsobilosti pracovníků NDT.

**Reference**

Vyhláška č. 358/2016 Sb. o požadavcích na zajišťování kvality a technické bezpečnosti a posouzení a prověřování shody vybraných zařízení.
NTD A.S.I. sekce VII
Interní dokumentace ČEZ a.s Divize jaderná energetika

# Potential of Active Thermography as a Non-Destructive Testing Method for Quality Assurance of Welded Joints

Eugen PRINTS [1], Igor KRYUKOV [1], Malte MUND [2], Christian ŠRAJBR [3], Klaus DILGER [2], Stefan BÖHM [1]
[1] Department for Cutting and Joining Manufacturing Processes, University of Kassel; Kassel, Germany;
Phone: +49 561 8043875; e-mail: e.prints@uni-kassel.de, i.kryukov@uni-kassel.de, s.boehm@uni-kassel.de
[2] Institute of Joining and Welding, TU-Braunschweig; Braunschweig, Germany;
e-mail: m.mund@tu-braunschweig.de, k.dilger@tu-braunschweig.de
[3] Edevis GmbH; Stuttgart, Germany; e-mail: christian.srajbr@edevis.de

**Abstract**

Welding is an integral part of the value chain in many areas of the manufacturing industry. A large number of irregularities can occur in the process. The evaluation of irregularities with regard to their admissibility is carried out on the basis of normatively defined evaluation groups (e.g. ISO 5817, ISO 10042). The basis for the evaluation of the weld or component quality are the results of non-destructive tests, which are used to determine the presence, extent and position of the weld irregularities. Active thermography is not yet sufficiently qualified for weld quality inspection, resulting in a standardization gap. The "QualiTWeld" research project is helping to establish thermographic testing for application to welded joints. This paper deals with the concepts for the reproducible production of test specimens. The potential of active thermography for testing welded joints is demonstrated by means of thermographic investigations on test specimens with realistic and idealized irregularities.

## 1. Introduction

Welding is particularly widespread as a joining process for metallic materials, belongs to the fourth main group in the classification of manufacturing processes, since it is possible to achieve a high-value material-to-material bond by applying force and/or heat. Welding can be carried out using various processes with and without filler metal/protective atmosphere, but the user is responsible for selecting a suitable joining process and ensuring that the required quality standards are met. To standardize the requirements for a welded joint, three evaluation groups are defined in standards (e.g. ISO 5817 or ISO 10042). These indicate the permissible limits of the irregularities that can occur in a welded joint. The classification into the evaluation groups is based on the type, location and geometric characteristics of the respective irregularity. Non-destructive testing methods (NDT) such as radiographic testing (RT), dye penetrant testing (PT) and ultrasonic testing (UT) are used to detect and quantify flaws. However, due to the physical principle of action, each testing method has advantages but also method-specific limitations, so that the holistic evaluation of a joint is often only possible through a combination of different NDT methods. [1–3]

Active thermography is not yet listed in the standard for evaluating welded joints, despite its robust and imaging results. The application of this non-destructive testing method is very diverse and includes the detection of imperfections occurring on the surface of a test specimen as well as those trapped inside. The suitability of active thermography for the detection of defects in different materials depends strongly on the type of excitation used. According to the functional principle of active thermography, heat currents are specifically generated in the test specimen, the propagation of which is influenced by the irregularities. The radiometric data recorded by the infrared camera can be used to reliably detect irregularities. It is the task of the tester to select a suitable source for the external energy input, the possibly necessary reference test specimen as well as the procedures for the evaluation of the measured data. The basic principles, equipment technology, evaluation procedures and idealized reference test specimens are covered in the standards for thermographic testing, but there is no uniform reference to the type, location and characteristics of the irregularities in the application-oriented evaluation groups from the field of welding technology. [4–8]

The aim of the current research project is to qualify the methods of active thermography for testing welded joints. The admissibility limits of relevant irregularities in the field of welding technology are recorded and compared with the detection limits of thermographic methods. For an objective evaluation of the detection capability of active thermography, test specimens are required that depict typical welding irregularities. In order to validate the thermography results, there is a need to have samples with the similar irregularities. Therefore, within the scope of the research project, methods for producing of the reproducible irregularities within the welded joints will be developed.

## 2. State of the art and target

The joining of metallic materials by welding is widely used in industrial production and is subject to high quality standards. Depending on the materials, joint type and component thickness, various welding processes can be used. When selecting the appropriate welding process for a specific joining task, the economic considerations play a decisive role in addition to functionality. [1] Due to the complexity of the joining process, a large number of irregularities can occur within a welded joint. A uniform classification of geometric irregularities on metallic materials during welding and related processes for fusion and pressure welding processes is given in ISO 6520.

For the evaluation of welded joints, the standardization differentiates between welding processes and the materials used. However, three evaluation groups B, C and D are uniformly defined, with group B presupposing the highest requirements for the welded joint. For fusion-welded joints in steel, nickel, titanium and their alloys, the permissible limits of irregularities for the three evaluation groups are summarized in ISO 5817. The definition of the evaluation groups of irregularities for arc welded joints on aluminum and its alloys is specified in ISO 10042. Process-specific standards apply to the evaluation of laser-beam welding processes and friction stir welding. The admissibility limits of individual evaluation groups for electron and laser-beam welded joints are summarized in ISO 13919 and differentiated by material in Parts 1 and 2. The guidelines for the evaluation of joints in friction stir welding are limited to the application on aluminum and its alloys and are laid down in ISO 25239-5. The designations of individual irregularities in the above-mentioned processes are introduced on the basis of ISO 6520. In general, a distinction is made between surface irregularities, internal irregularities, irregularities in the seam geometry and multiple irregularities. [2, 9–12]

The cracks with the order number 100 according to ISO 6520-1 are considered to be not permissible for all three evaluation groups under surface irregularities and internal irregularities. Lacks of fusion (order numbers 401/402) also belong to the critical irregularities, since they are not permissible as surface irregularities and as internal irregularities for welding either steel or aluminum alloys according to evaluation group B. The volumetric irregularities (e.g. pores, order number 2011) are partially tolerated within the evaluation groups in compliance with the specified permissibility limits. The limits are defined as a function of the nominal weld thickness. [2, 9–13]

According to the welding standardization described above, critical imperfections can be divided into two groups: "impermissible" and "conditionally permissible". The first group includes, for example, cracks and bonding defects, the occurrence of which must be detected in welded joints. Conditionally permissible are irregularities whose presence is tolerated on the basis of certain limit values (e.g. pores). The permissible size of an irregularity depends on the materials used, the type of joint and the component wall thickness. Based on normative specifications, impermissible weld irregularities can be detected with the aid of NDT procedures. The NDT methods established in the field of welding

technology include VT, PT, MT, RT, UT and ET. However, process-specific limitations of NDT methods with regard to detection capability are to be expected in practice due to the type, location and characteristics of an irregularity. [3]
By qualifying active thermography for the testing of welded joints, the established test methods can be supplemented by another reliable NDT method for detecting relevant irregularities.

## 3. Results

The production of test specimens with defined irregularities is an essential aspect for the qualification of thermographic processes with reference to the normative specifications from the field of welding technology. Within the scope of the project, concepts are being developed that enable the reproducible insertion of relevant flaws while specifying their type, size and position. The selection of suitable concepts is carried out iteratively under permanent control by destructive and non-destructive testing methods, for example metallographic examinations and radiographic testing.
In welding standards, the quality requirements are differentiated on the basis of the type of material and the joining process used, since, on the one hand, the type of imperfection can arise process-specifically and, on the other hand, the quality of the joined joint must be evaluated with certain defects depending on the material for the application. The qualification of thermographic processes for welded joints is to be carried out in a user-oriented manner within the scope of the project. For this reason, manual arc welding as a conventional fusion welding process, laser beam welding as a beam welding process and friction stir welding as a special welding process, which belongs to the pressure welding processes, are used for test specimen production. An aluminum alloy (EN AW 6082-T6), an unalloyed structural steel (S235JR) and an austenitic chromium-nickel steel (X5CrNi18-10) were selected as representative materials for the test specimens because they are widely used in industrial applications. [14, 15]
Realistic irregularities can be produced by varying the welding system parameters, changing the tool parameters and manipulating the tool guidance or clamping device on mechanically controlled welding systems. These approaches reproduce real problems in production, such as tool wear, errors in programming or deviation of the welding parameters as well as incorrect positioning of the workpieces.
For an objective evaluation of active thermography methods, it is necessary to prove the detection capability of real irregularities and to compare the results with established NDT methods. Active thermography implies the use of an external excitation source, with the help of which heat flows are induced in the test specimen. The irregularities occurring inside or on the surface disturb the heat transport process. The irregularities of the heat flow can be detected by means of the infrared radiation (IR) emitted from the surface of the test specimen by means of an IR camera. The heat input is usually via conduction, thermal radiation or induction. Furthermore, the choice of the positioning of the excitation source, the IR camera and the test object is differentiated into transmission and reflection arrangement. [6] For a comparison of different excitation sources with respect to the detection of impermissible and conditionally permissible irregularities, reference is made here to previous work. [16]
The test specimens presented in the following were examined using induction-excited thermography in transmission and reflection configuration. An actively cooled IR camera SC5600-M with an InSb photo quantum detector was used as the recording medium. Various inductors were used for external heat input. The evaluation of thermographic measurements is limited to the methodology of classical pulse-phase thermography. [17]

### *3.1 Lack of fusion and cracks*

*3.1.1 Lack of fusion and cracks (real)*

Lack of fusion is one of the critical irregularities in welding, its occurence is associated with a strong reduction in the strength of a joint. Figure 1 shows the results of investigations of a laser-welded lap joint (150 x 50 mm each with 1.5 mm wall thickness, S235JR). Lack of fusion was produced in the test specimen by manipulating the fixture at a laser power of 750 W, welding speed 2 m/min and a focus position of 0 mm. A defined increase in the size of the joint gap leads to the formation of the so-called "false friend", in which there is no material bond between the joining partners. During the thermographic test, heat was introduced in the lower sheet at an induction frequency of 12.5 kHz and pulse width modulation (PWM) of 700 ‰, which represents a performance parameter with itself, within the pulse duration of 0.1 s with the help of a stepped coaxial inductor. The heat flow toward the test specimen surface is disturbed by material separation between the joining partners. This change in heat flow can be detected by radiometric data from an IR camera. The area of the lack of fusion stands out clearly in the phase representation (Figure 1, 2a) from areas where a material-locked joint is present. The graphical representation of the phase position along the blue evaluation line (2b) illustrates that the heat buildup caused by the lack of fusion leads to a decrease in phase position. The results of thermographic testing were confirmed by X-ray examination of the specimen. The sectional views along (1a) and at the marked points orthogonal to the welded joint (1b-1d) illustrate the extent of the irregularity introduced.

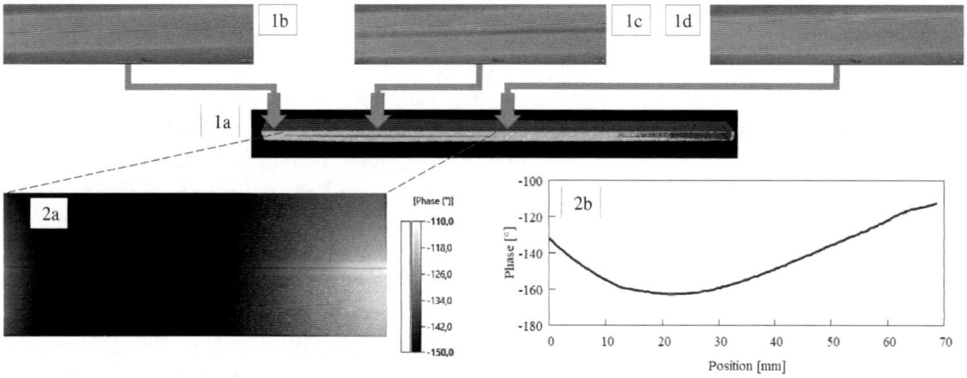


**Figure 1. Radiographic inspection: sectional plane along a laser beam weld joint (1a), sectional plane transverse to the weld joint (1b-d). Induction-excited thermography: phase representation at 0.05 Hz (2a), phase progression along weld center at 0.05 Hz (2b)**

In addition to lack of fusion in welded joints, cracks are also one of the critical irregularities in the welding of metallic materials. Figure 2 shows a welded joint made by manual arc welding (butt joint, 220 x 100 mm each with 8 mm wall thickness, S235JR). Due to the metallurgical incompatibility of the molten materials, the addition of copper to the weld metal results in solidification cracks. By visual inspection, clearly pronounced cracks running transverse to the direction of welding can be detected (see Figure 2, 1). By means of dye penetrant inspection, further cracks with smaller dimensions can be detected (see Figure 2, 2). The dye penetrant is carried to the surface by the developer, and this can cover up smaller indications during the test, so that their detection is not possible. With the help of active thermography, a higher detection probability can be achieved than with dye penetrant testing. Figure 2 (3) shows the result of thermographic testing in reflection configuration. Using the inductor specially developed for crack detection, eddy currents were applied close to the surface at 60 kHz excitation frequency with a PWM of 540 ‰ for a pulse duration of 0.1 s. The test specimen was then subjected to a PWM of 540 ‰. For the test, the specimen was painted black to increase the emissivity. Based on the amplitude display at an evaluation frequency of 10 kHz, which corresponds to a test duration of 0.1 s, the cracks can be clearly visualized. In addition to the short test duration, the high automation potential of thermographic methods is another advantage of this testing technique. In addition to the detection limits of active thermography, suitable evaluation methods will be developed within the scope of the project in order to minimize the specific disadvantages, such as the dependence on the surface condition.

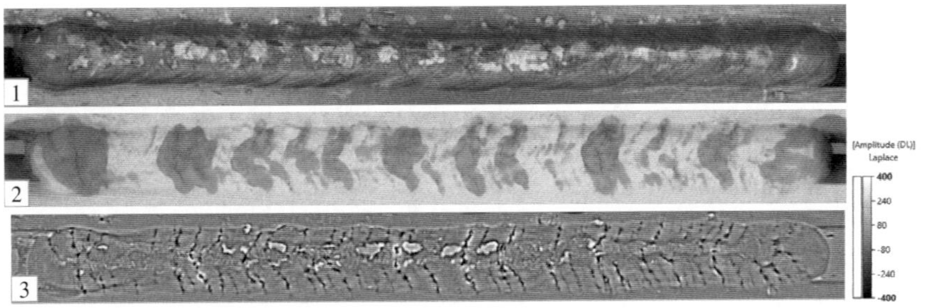


**Figure 2. Cracks within a manual arc weld joint. Photo of the welded joint (1), dye penetrant inspection (2), induction-excited thermography [amplitude display with edge visualization by Laplace filter (3x3) at an evaluation frequency of 10 Hz] (3)**

### *3.1.2 Lack of fusion and cracks (idealized)*

To record the detection limits of thermographic methods for the inspection of welded joints, the location and extent of defects must be introduced in a controlled manner during test specimen production. Figure 3 (1) schematically presents a concept for butt joints that allows the position and size of a crack-like flaw within the test specimen to be precisely defined by several weld beads using the friction stir welding method. Here, a weld with a shortened pin is first performed along the butt edge to map defined crack depths. The crack length is generated in a controlled manner by further welds with a reference pin length orthogonal to this. In this way, the reference to the acceptance limits can be ensured by evaluation groups from the field of welding.

For the specimen preparation two plates of Aluminium (EN AW 6082-T6) with 3 mm wall thickness were joined by friction stir welding. The spacing of the four transverse welds was selected so that three cracks of equal length were produced. The definition of the crack length was gradually increased from 0.5 mm to 5 mm (cf. Figure 3, 1). Visually, cracks generated in this way are not visible up to a length of 4 mm, but with the aid of dye penetrant testing, idealized cracks from a length of 1.5 mm could be clearly detected (cf. Figure 3, 2). A characteristic feature of this type of testing is the increased leakage of the dye penetrant at the beginning and end of the crack.

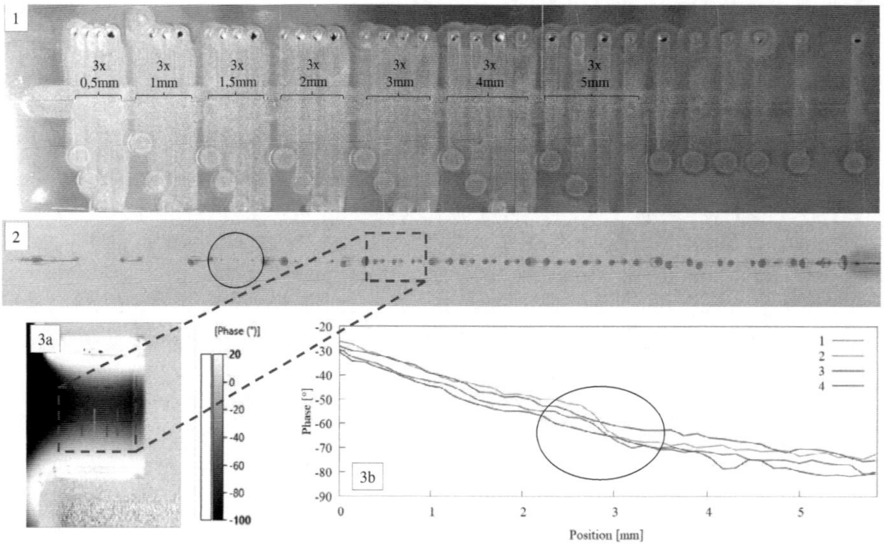


**Figure 3. Investigation of the concept of "idealized cracks". Side of the test specimen facing away from the weld seam (1), dye penetrant test (2). Induction-excited thermography: phase representation at evaluation frequency 0.1 Hz (3a), phase progression along the marked areas (3b)**

The specimen was subjected to induction-excited thermographic testing at excitation frequency of 30 kHz, PWM of 700 ‰ and pulse duration of 0.5 s in reflection configuration was operated. Figure 3 (3a) presents a phase plot at an evaluation frequency of 0.1 Hz within the ROI (region of interest). The generated cracks from a length of 3 mm could be clearly detected. For clarity, signal traces along marked regions are shown graphically in Figure 3 (3b). The introduced heat is hindered in its propagation by the idealized crack. This difference compared to the fully welded areas can be seen by the dip in the phase angle. Heat flow thermography is only suitable for crack detection to a limited extent, since the detection capability depends strongly on the position and depth, and the duration of the investigation (10 s in this example) is quite long. The aim of the project is to select or develop suitable excitation technology and optimum evaluation methods to be able to determine the detection limits of active thermography for testing welded joints.

**If the production step with transverse welds is omitted during specimen production (cf. test specimen [Figure 3, 1]), lack of fusion occurs along the complete butt edge, which is considered impermissible according to ISO 10042 (ordinal number 401). Thermographic tests were performed with the same hardware and test parameters as presented in Figure 3. The suitability of heat flow thermography for detecting lack of fusion in the butt joint can be seen from the phase diagram presented in Figure 4 (a and b) and the dip in the phase position in the signal curve. The lack of fusion with a depth of 1.1 mm could be clearly detected thermographically on the basis of its effect on heat propagation.**

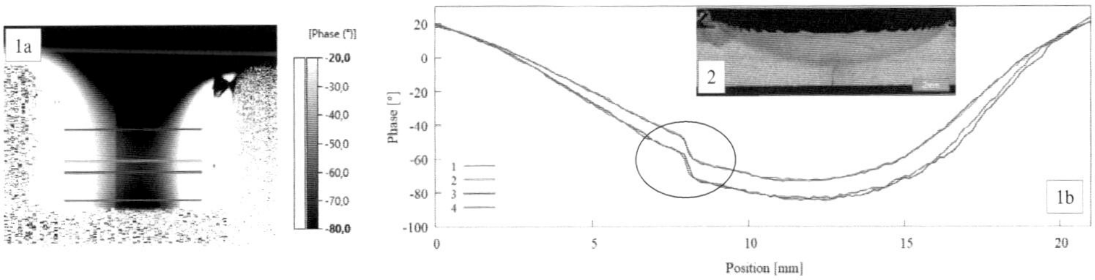


**Figure 4. Investigation of the "lack of fusion" concept. Side of the test specimen facing away from the weld seam, induction-excited thermography: phase representation at evaluation frequency 0.1 Hz (1a), phase progression along the marked areas at evaluation frequency 0.1 Hz (1b). Micrograph of weld seam cross section (2)**

### *3.2 Cavities*

*3.2.1 Cavities (real)*

According to normative specifications in the field of welding technology, voids are to be classified as "conditionally permissible", since their occurrence is partially accepted depending on their size and wall thickness of the joint. Figure 6 shows a comparison of the inductively excited thermographic test and the X-ray reference test of a laser-beam welded lap joint (150 x 50 mm each with 1.5 mm wall thickness, S235JR). Increased laser power (1000 W) and a focus position of - 2 mm at a welding speed of 2 m/min were used to fabricate the welded specimen. The selected welding parameter configuration results in irregulary pores in the joint gap.

The hardware and the excitation parameters when performing thermographic testing correspond to those already mentioned in Figure 1. The heat inductively introduced in the lower joint partner is disturbed in its propagation by the irregularities within the weld joint. In the phase diagram (see Figure 5, 1b), inhomogeneities in the phase position can be seen in the selected area of the welded joint. The irregularities are assigned by comparing the graphical representation of the phase position along the blue evaluation line (Figure 5, 1a) with the results of the radiographic test. Here, the marked area with red dotted lines within the thermographic evaluation of the ROI corresponds to the depicted evaluation of the radiographic inspection (cf. Figures 2a and 2b). The shape and position of the cavity affects both the phase progression and the phase position. For example, the cavity marked with the number 3 can be recognized by its deepest phase position due to its size. The single pore (No. 2), with a diameter of about 0.8 mm, also stands out clearly in the phase progression due to a local minimum. The irregularities, marked with the numbers 1 and 4, each consist of several single pores and are detectable in the phase progression by decreasing phase position. The slope of the signal curve can be used to make a qualitative estimate of the extent of the irregularities (cf. border areas of the ROI in the evaluation of the thermographic and transmission tests).

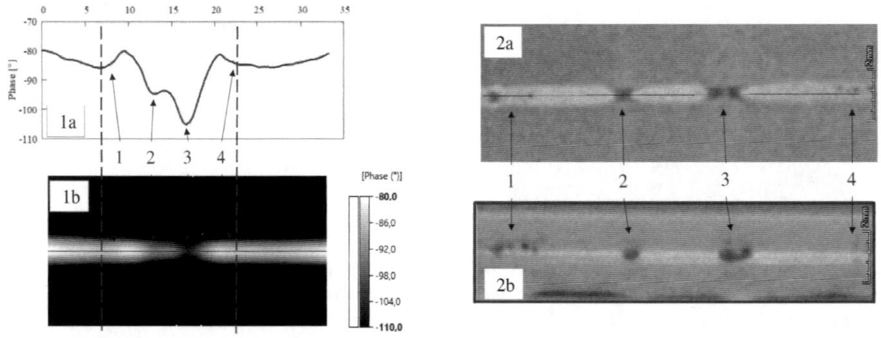

**Figure 5. Induction-excited thermography: phase progression along weld center line(1a), phase representation at 0.5 Hz (1b). Radiographic inspection: top view, sectional plane between the individual sheets (2a), sectional plane orthogonal to them (2b)**

*3.2.2 Cavities (idealized)*

In addition to the controlled generation of bonding defects and cracks, concepts are being developed within the scope of the research project which allow the targeted introduction of volumetric irregularities. Based on the results presented in Figure 6, a concept using workpiece preparation is discussed. In the first step, 6 holes with diameters ranging from 1 mm to 5 mm were drilled 4 mm deep at defined intervals in a plate made of EN AW 6082-T6 with a wall thickness of 5 mm. In the next step, friction stir welding was performed along the hole centers in a lap joint, with a sheet made of EN AW 6082-T6 with 1.5 mm wall thickness serving as the upper joining partner (see Figure 6, 1).

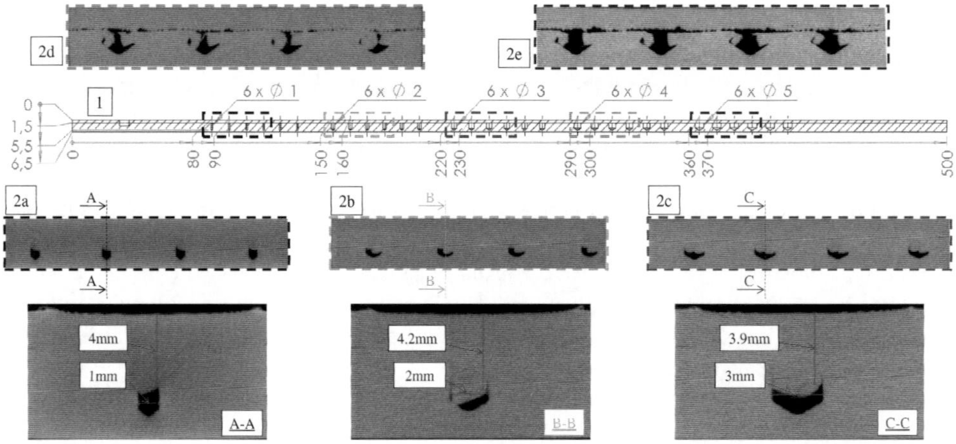

**Figure 6. Test specimen for the generation of cavities. CAD drawing, sectional view along the weld joint (1). Results of radiographic inspection: center plane along weld joint (2a-e), plane transverse to weld joint at selected holes (A-A, B-B, C-C)**

In this concept, reproducibility of the insertion of voids by friction stir welding was investigated. Radiographic investigation was performed using a µCT system (Zeiss xradia 520 Versa). Due to the higher spatial resolution, the measuring field was selected in such a way that 4 of 6 holes could be imaged with one measurement. Due to a combination of translational and rotational motion of the friction stir welding tool, the plasticized base material is introduced into the hole (cf. Figure 6, 2a to 2e), Defects generated using hole diameter 1 mm can be approximately considered as volumetric irregularities. As the hole diameter increases, the cavity is filled with the base material, so that the ratio of cavity diameter to height of 1:1 is not given (cf. Figures 2b and 2c with corresponding planes transverse to the welding direction). When using hole diameters larger than 4 mm (cf. Figs. 2d and 2e), the lack of base material in the stirring zone leads to lack of fusion between the joining partners in some areas. The reproducibility of the defect introduction with the help of the investigated concept is given. In order to counteract the negative effects of varying the defect size described above, the friction stir welding tools will be further developed and the welding parameters optimized in the further course of the project.

The specimen presented in Figure 6 was investigated using induction-excited thermography in transmission configuration. Heat was introduced from the side away from the weld using the induction frequency of 12.5 kHz and the pulse width modulation of 735 ‰ over a pulse duration of 0.5 s through a coaxial stage inductor. The weld surface was painted black to increase the emissivity. In each case, the evaluation was limited to the 4 holes examined radiographically. Within the selected areas, the heat input by induction is almost homogeneous. Figure 7 presents the evaluation of the thermographic investigations. Both the phase diagram (cf. 1a-1e) and the corresponding signal curves (2) show that bore sizes of up to 2 mm (cf. diagram 1b) can be unambiguously detected. Thereby, hole depth is more than twice as large compared to hole diameter. The heat accumulation caused by a cavity is manifested by lower phasing compared to the fully welded areas. The irregularity with a diameter of 1 mm could not be detected; the ratio of hole depth to hole diameter corresponds to a factor of 4. The irregularities with hole diameters of 4 mm and 5 mm have a deeper phase (see Fig. 2), which can be attributed to lack of fusion between the joining partners (see Fig. 6, 2d and 2e, and Fig. 7, 2).

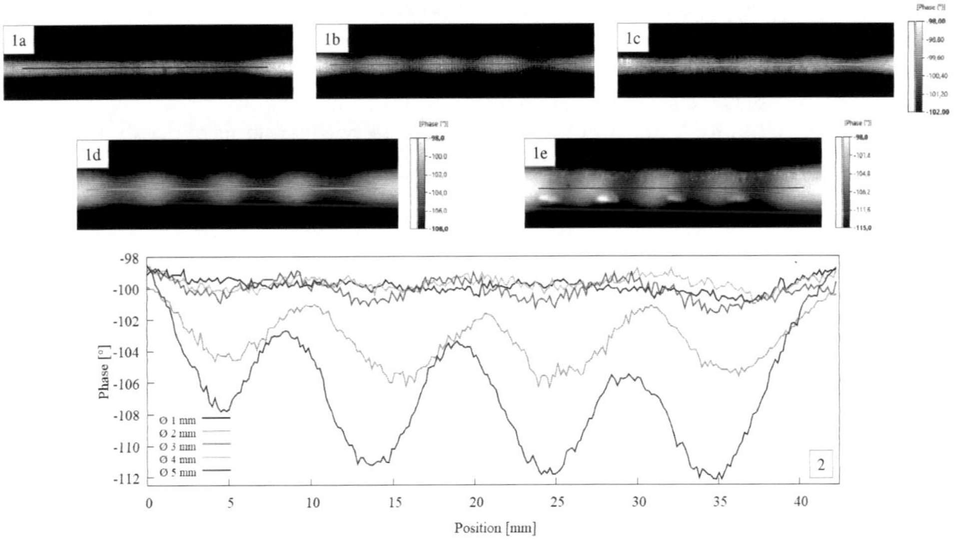

**Figure 7. Investigation of the "cavities" concept. Induction-excited thermography: phase representation at an output frequency of 0.6 Hz (1a to 1e), phase progression along the marked areas (2)**

Within the scope of the project, further concepts for the production of volumetric irregularities will be pursued in order to reduce the degree of idealization and to strengthen the reference to the normative specifications from the field of welding technology.

## 4. Conclusions

the methodological approach for qualifying active thermography as a non-destructive testing method for testing welded joints was investigated. The development of concepts for production of reproducible specimen with defined irregularities was based on the normative specifications from the field of welding technology. Emphasis was placed on the mapping of admissibility limits for various irregularities as a function of the welding processes and materials that are widely used in industrial production. These test specimens are necessary to determine the detection limits of active thermography and to integrate the method into the series of established non-destructive testing methods for quality assurance of welded joints.

The thermographic investigations presented here, demonstrate the potential of active thermography for testing welded joints. Lack of fusion of an overlap welded laser beam joint as well as pores deposited within the welded joint could be clearly detected by induction excited thermography in transmission order. The cracks generated during manual arc welding could also be detected by induction-excited thermography. The results of thermographic investigation were validated by radiographic and dye penetrant testing.

Test specimens with idealized irregularities are necessary for an objective evaluation of the detection limits of active thermography methods. Thus, the influence or limitation of the thermographic test can be determined by the welding process and the material. This approach was explained using the exholeample of test specimens with idealized lack of fusion, cracks and cavities produced by friction stir welding. Results of thermographic examinations were presented and discussed.

The conceptual design of the test specimens with realistic and idealized irregularities will be continuously developed within the research project. The suitability of various excitation sources for testing welded joints is being investigated and the detection limits of active thermography are being evaluated in relation to the permissibility limits of irregularities relevant to production in the field of welding technology.

**Acknowlegements**

The investigations were carried out as part of the QualiTWeld project (Ref. No. 03TN0003), which is funded by the Federal Ministry for Economic Affairs and Energy as part of the "WIPANO - Knowledge and Technology Transfer through Patents and Standards" program. The authors would like to thank the project partners for their cooperation. The responsibility for the content of this publication lies with the authors.

## References

1. Matthes K-J, Schneider W, Werner (Hg), 'Schweißtechnik. Schweißen von metallischen Konstruktionswerkstoffen', 6. Aufl., München, Leipzig im Carl Hanser Verlag, 2016.

2. German Institute for Standardization ISO 5817: 'Welding - Fusion-welded joints in steel, nickel, titanium and their alloys (beam welding excluded) - Quality levels for imperfections', June 2014.

3. German Institute for Standardization ISO 17635: 'Non-destructive testing of welds - General rules for metallic materials', August 2010.

4. Maierhofer C, Busse G, Holtmann N, et al., 'Standardisierung der Lockin - Thermografie: Vorstellung des TNS-Projektes und Auswahl der Probekörper', Thermographie-Kolloquium, 2013.

5. Frey J, Srajbr C, 'Vollautomatische Thermografie ersetzt Magnetpulverprüfung an gehärteten Werkstücken', DGZfP-Jahrestagung, 2018.

6. German Institute for Standardization DIN EN 17119, 'Non-destructive testing - Thermographic testing - Active thermography', October 2018.

7. German Institute for Standardization DIN 54183, 'Non-destructive testing - Thermographic testing - Eddy-current excited thermography', February 2018.

8. German Institute for Standardization DIN 54184, 'Non-destructive testing - Pulse thermography using optical excitation', October 2017.

9. German Institute for Standardization ISO 10042, 'Welding - Arc-welded joints in aluminium and its alloys - Quality levels for imperfections', February 2006.

10. German Institute for Standardization ISO 13919-1, 'Electron and laser-beam welded joints - Requirements and recommendations on quality levels for imperfections - Part 1: Steel, nickel, titanium and their alloys', March 2020.

11. German Institute for Standardization ISO 13919-2, 'Electron and laser-beam welded joints - Requirements and recommendations on quality levels for imperfections - Part 2: Aluminium, magnesium and their alloys and pure copper', May 2020.

12. German Institute for Standardization 25239-5, 'Friction stir welding - Aluminium - Part 5: Quality and inspection requirements', December 2020.

13. German Institute for Standardization ISO 6520-1, 'Welding and allied processes - Classification of geometric imperfections in metallic materials - Part 1: Fusion welding', November 2007.

14. Dilthey U., 'Schweißtechnische Fertigungsverfahren 1. Schweiß- und Schneidtechnologien' Bd. 1. VDI-Buch, 3. Aufl. Springer-Verlag, 2006.

15. Dilthey U., ' Schweißtechnische Fertigungsverfahren 2. Verhalten der Werkstoffe beim Schweißen', Bd. 2. VDI-Buch, 3. Aufl. Springer-Verlag, 2006.

16. Malte Mund, Eugen Prints, Igor Kryukov, Jonas Hensel, Stefan Böhm, Klaus Dilger. Qualifizierung der aktiven Thermografie zur Prüfung von Schweißverbindungen an Prüfkörpern mit realitätsnahen Unregelmäßigkeiten', DGZfP-Jahrestagung, 2021.

17. Maldague XP, Marinetti S., 'Pulse phase infrared thermography', Journal of Applied Physics 1996; 79: 2694, doi:10.1063/1.362662, 1996.

# NEDESTRUKTIVNÍ TESTOVÁNÍ HLAV VÁLCŮ, SEDEL A VENTILŮ VE VÝROBĚ A SERVISU

# NON-DESTRUCTIVE TESTING OF CYLINDER HEADS, SEATS AND VALVES IN PRODUCTION AND SERVICE

**Břetislav SKRBEK[1], Jakub MRÁZ[1]**

[1] Technical university of Liberec, Studentská 2, Liberec 461 17, bretislav.skrbek@tul.cz

***Abstrakt***

*Nedestruktivní testování užitných vlastností dílů kolem spalovacího prostoru. Měření tuhosti a pevnosti hlav válců – kombinované nedestruktivní měření. Predikce opotřebení sedel ventilů z legované litiny nebo tvrdokovu typu stellit. Stanovení kritické meze degradace ocelí hlav výfukových ventilů, kumulace teplotního přetížení – roste křehkost. Ultrazvukové a magnetické testování. Zkušenosti, příklad technologického postupu. Průmysl 04, roste konkurence schopnost i firmě s menší produkcí.*

***Klíč. slova:*** *Součásti motorů; nedestruktivní testování; užitné vlastnosti; výroba, servis.*

***Abstract***

*Non-destructive testing of utility properties of parts around the combustion chamber. Measurement of stiffness and strength of cylinder heads - combined non-destructive measurement. Prediction of wear of valve seats made of alloyed cast iron or carbide type stellite. Determination of critical limit of degradation of steels of exhaust valve heads, accumulation of thermal overload - increases brittleness. Ultrasonic and magnetic testing. Experience, an example of a technological process. Industry 04, growing competitiveness even for a company with smaller production.*

***Key words:*** *Engine parts; non-destructive testing; useful properties; production; service.*

## 1. Úvod

Spalovací pístové plynové motory mají stále perspektivu i v nastupující vlně elektrizace pohonů zejména u dopravních prostředků. Bezkonkurenční využití pístových plynových motorů je a bude v kogeneračních jednotkách, které ekologicky využívají plyny ze skládek odpadů, zemědělských procesů, z lokálních vzdálených zdrojů, jako záložní zdroje tepla a elektřiny. Funkčnost a životnost dílů obklopujících spalovací prostor (obr.1) limituje ekologii a užitnou hodnotu motoru. Tvoří ji materiál a geometrie hlavy válců, vedení a sedla ventilů, ventily, píst a vložený válec. K sledování shody a degradace materiálu těchto dílů lze úspěšně využívat metod nedestruktivního testování. Přispívá k uplatnění prvků průmysl u 0.4. Tento příspěvek se zabývá detaily 1 až 4 z obr.1.

## 2. Hlavy válců

Vysoká tuhost zejména desky ventilů hlav válců zaručuje rovnoměrné a stabilní sezení sedlových ploch ventilů v sedlech ventilů po montáži (stažení hlavovými šrouby k bloku válců) a za proměnných provozních sil. Vysoká hodnota tepelné vodivosti umožňuje účinný odvod tepla deskou ventilů do chladícího media. Pro pístové motory o výkonu nad 100kW se aplikují litinové hlavy. Nejvyšší tepelnou vodivost a nejnižší tuhost a cenu mají litiny s lupínkovým grafitem. V praxi se uplatňuje škála jakostí EN GJL 250 až 350. Z hlediska materiálových vlastností optimální litiny s červíkovým grafitem např. EN GJV 350 komerční slévárny nedokážou za přijatelnou cenu vyrobit.

Tuhost vyjádřená hodnotou počátečního modulu pružnosti $E_0$ litinových hlav v rámci přejímky se kontroluje ultrazvukem dle jednoduchého vztahu

$E_0 = (L/L_u * K)^2$ [MPa] (1)

L... skutečný rozměr (výška hlavy v místě středového nálitku pro otvor hlavového šroubu).

$L_u$..ultrazvukem (seřízen na podélnou rychlost oceli $c_L = 5920$ m/) změřená výška L.
K...konstanta zahrnující měrnou hmotnost a příčnou rychlost $c_T$.

Konstanta K se stanoví měřením rozměrů, hmotnosti a ultrazvukem na štíhlých válcových vzorcích [1]. V běžném provozu s dodávkou odlitků hlav se obdrží atest o chemickém složení a dalších vlastnostech dle TDP a záložní zkušební tyč D30-250 mm. Po jejím ohrubování na D 28 – 210 mm se změří aktuální hodnota K [2]. Využívá se velkých ( d20 mm) jednoduchých či dvojitých přímých sond 2MHz.

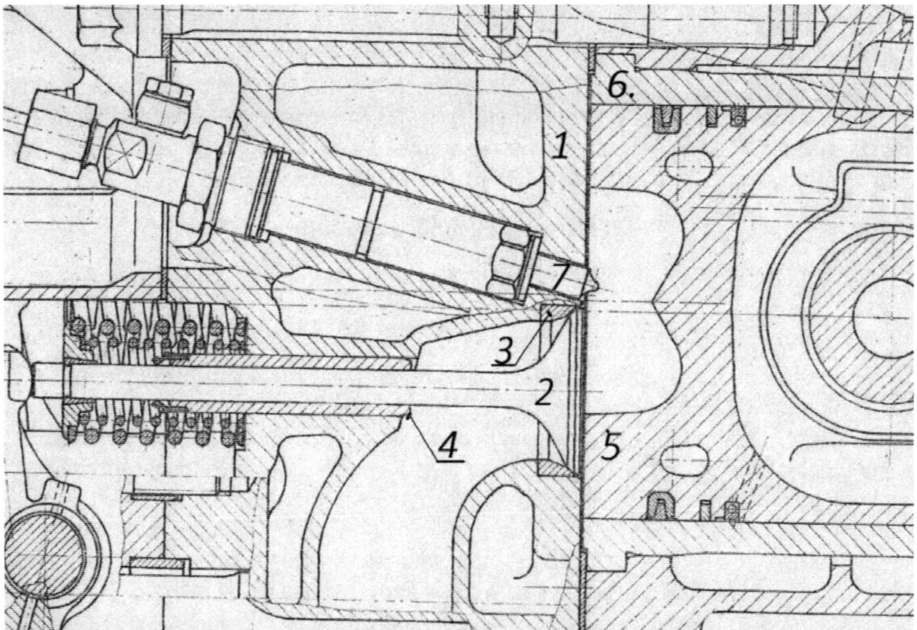

Obr.1. Část řezu motorem – okolí spalovacího prostoru. 1 – hlava válců; 2 – ventil; sedlo ventilu; 4 – vedení ventilu; 5 – píst s pístními kroužky; 6 – vložený válec; 7 – tryska (u vznětového motoru) nebo svíčka s tryskou u plynových motorů.

Vlastní přejímání odlitků spočívá v kombinovaném měření ultrazvukem a magnetickým tvrdoměrem DOMENA přes neupravený základovou barvou natřený povrch. Sondou 2 MHz se prozvučí ($L_u$) nálitek středního šroubu, kterým se upevňuje k bloku válců, celá výška L = 125mm hlavy. Snímač DOMENy se přiloží mezi sedlovými otvory na spodní desku hlavy. Změřená hodnota M [A/m] intenzity zbytkového magnetického pole je v lineárním vztahu k tvrdosti HB dle Brinella.

$$HB = 0{,}6 \cdot M + 105 \qquad (2)$$

Řídící počítač strukturoskopu TELIT stanoví i pevnost litiny v měřeném objemu materiálu.

$$R_m = a \cdot (L/Lu)^b \cdot HB^c \qquad [MPa] \qquad (3)$$

Každý odlitek má předlit datum a pořadové číslo. V případě neshody některé vlastnosti tak lze dohledat příčiny ve slévárně nejen u nových odlitků, ale i u z provozu reklamovaných hlav válců. Příklad grafického vyjádření vlastností souboru 50ti dodaných hlav ukazuje obr. 2.

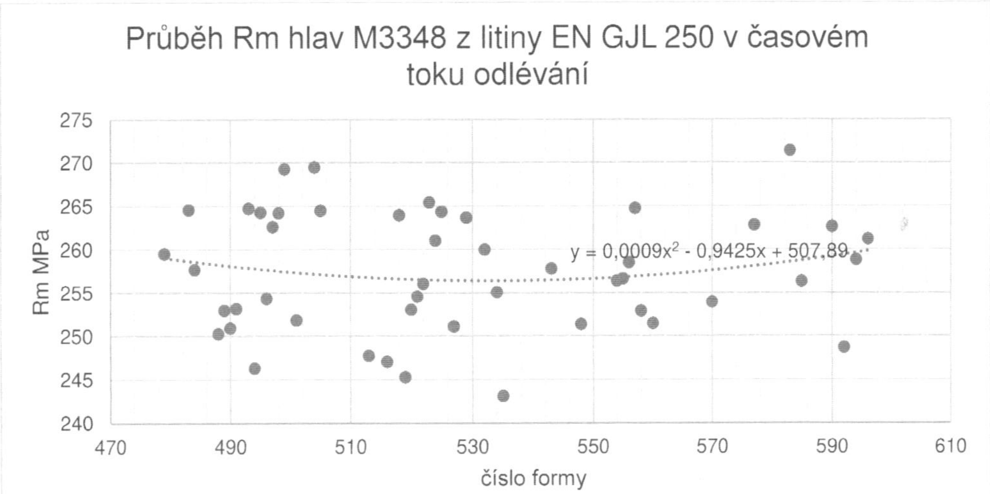


*Obr.2.*

### 3. Litiny sedel a vodítek ventilů

Monopolním dodavatelem sedel a pístních kroužků od r. 1932 je v českých zemích slévárna a obrobna v Komárově u Hořovic. Odolnost struktury vůči opotřebení spočívá ve vytvoření stabilní (v rozsahu pracovních teplot) perlitické či sorbitické matrice a skeletu otěruvzdorných fosfidů a karbidů. Grafit podporuje tepelnou vodivost a snižuje koeficient tření. Složení (střední hodnoty) základních litin s lupínkovým grafitem na pístní kroužky (B0 až B9), sedla (BS4 až BY) a vedení (V1, V2) slévárny „BUZULUK“ Komárov [3] předkládají interní normy KOMA. Tolerance obsahu základních prvků C, Si, P jsou poměrně velké. Nejsou zde vázány stupni eutektičnosti SE, které souvisí s důležitou hodnotou E. Pro litinu s lupínkovým grafitem litou do písku a stěnu L 15mm lze z českých norem [2] sestavit tabulku 1.

*Tab.1. Střední hodnoty SE a E jakostí LLG dle ČSN.*

| ČSN 4224++ | 10 | 15 | 20 | 25 | 30 |
|---|---|---|---|---|---|
| SE | 1,08 | 1,02 | 0,96 | 0,9 | 0,84 |
| $E_{oL15}$ * 0,01 [GPa] | 0,889 | 0,973 | 1,099 | 1,25 | 1,44 |

Kroužky sedel se odlévají každý zvlášť do keramické formy. Vysoké obsahy Mo a Cr variant BX a BY působí vznik martenzito – bainitické matrice při rychlosti ochlazování přirozeným chladnutím ve formě. Proto před obráběním nutno odlitky stabilizovat popouštěním na teplotu vyšší než provozní. Sedla BX s BY se popouští na 570 až 580 $^0$C. Výsledná matrice pak obsahuje i sorbit o rozsahu dovolené tvrdosti 31 – 40HRC 1mm pod povrchem.

Sedla z BX se běžně používají na přeplňované vznětové motory. Litina BY byla vyvinuta pro motory na bezolovnatý benzin. Na plynových zážehových motorech vykazovala ale vyšší opotřebení (4). Opotřebení A při konkrétním způsobu a délce provozu sedel lze s vysokou pravděpodobností (R=0,91) předpovědět nedestruktivně změřenými vlastnostmi [1] – tvrdostí H dle HRC a modulem pružnosti.Eo před montáží sedla výrazem (4). Samotný modul pružnosti E (počáteční – změřený ultrazvukem) se spočítá podle (2).

$$A = 7{,}6 - 1{,}37 \times H \times E \times 10^{-3} \qquad [mm] \qquad (4)$$

V rámci nedestruktivní přejímky musí hodnota $A_{max}$ = 0. Nyní se však pro vysoce namáhané motory používají sedla z tvrdokovů typu Stellit, intermetalických sloučenin či produktů práškové metalurgie. Vedení ventilů motorů LIAZ a TEDOM prošlo dlouhým vývojem materiálů [5, 6], který se ustálil na optimu životnosti a ceny – izotermicky kalené laciné nelegované litině s lupínkovým grafitem (předpis jen SE a HB). Technologie izotermického kalení má úzké procesní okno (malá tolerance teplot a výdrží). Nutno aplikovat nedestruktivní (MT, UT) kontrolu [4].

## 4. Ventily

Ventily pístových spalovací motorů přesně vymezují režim sání, spalování hořlavé směsi a výfuk spalin. Na přesném vedení ventilů závisí výkonová stabilita a ekologie provozu motoru. Ventil tvoří dřík z chromové martenzitické oceli (např. ČSN 41 7115) s kalenou stopkou, který je třecím svarem spojen s hlavou ventilu. Sklon sedlové plochy ventilů bývá 45$^0$.Pro plynové motory 30 $^0$ Austenitická ocel hlavy ventilu (např. ČSN 41 7465) se v sedlové ploše musí "opancéřovat" Co karbidickým návarem P37 (STELLIT F). Většina provozních selhání ventilů únavovým lomem souvisela s výskytem nadkritických nespojitostí v místě začátku a současně konce obvodového návaru sedlové plochy. Kvalitu třecího svaru a homogenitu návaru se před 30ti lety ještě kontrolovala uživateli ventilů [8, 9]. Kvalitu třecího svaru a homogenitu návaru kontroluje ultrazvukem nyní výrobce ventilů (TRW ap). Austenitická ocel hlavy ventilu nového ventilu je paramagnetická. Dlouhodobým teplotním přetěžováním se však tvoří složky struktury, které působí křehkost (pokles hodnot KCU) ale i feromagnetické chování [7,10,11]. Tohoto jevu se využilo k servisním kontrolám provozovaných ventilů ještě zamontovaných v hlavách válců. Příklad technologického postupu kontroly je v příloze.

## 5. Závěr

Nástroje a metody klasického nedestruktivního zkoušení nespojitostí i nedestruktivní strukturoskopie výrazně přispívají k vyšší jakosti finálního výrobku – spalovacího pístového motoru a zkvalitnění servisních služeb. Přístroje zabezpečující uvedená zkoušení předkládají kvantitativní data, která elektronickým přenosem do firemní sítě uplatňují smysl průmyslu 0.4.

Publikace byla podpořena Katedrou materiálu SF TU v Liberci z konta 117 pro vědu a výzkum.

**Literatura:**

[1] Obraz J. Ultrazvuk v měřící technice. SNTL Praha.1989.

[2] Skrbek. Mráz. Rozšířené využití zkušebních tyčí atestů odlitků. 56. Slévárenské dny, Brno 2019, sborník s.192 – 195, ISBN 978-80-02-2882-6.

[3] KOMA 42 0060 až 0090. Interní normy Slévárny pístních kroužků a sedel ventilů Komárov.

[4] ANDRŠOVÁ, Z. Nedestruktivní strukturoskopie izotermicky kalených litin. Disertační práce, Technická univerzita v Liberci, 2014.

[5] KAPOUNOVÁ, P. Vliv tepelného zpracování na mezní ohybové namáhání materiálu ventilových vedení. Diplomová práce KMT 139, TU v Liberci 2001.

[6] SKRBEK, B., Mráz, J. Technická přejímka důležitých odlitků produktu. 56. Slévárenské dny, 2019, Brno.

[7] Fiala, J., Kolega, M. Strukturní projevy provozní expozice austenitických ventilových ocelí. Sborník conference Provozní degradace austenitických ventilových ocelí, Liberec 2006, s. 24 – 39, ISBN 80-7372-113-9.

[8] Eckhardth, E. Detekcia trhlín v trecom zvare drieku ventilu. Sborník conference Provozní degradace austenitických ventilových ocelí, Liberec 2006, s. 49 – 53, ISBN 80-7372-113-9.

[9] Regazzo, R., Regazzová, M. Ultrazvuková diagnostika výfukových ventilů dieslových motorů. Sborník conference Provozní degradace austenitických ventilových ocelí, Liberec 2006, s. 92 –95, ISBN 80-7372-113-9.

[10] Králík, M. Provozní změny struktury a fyzikálních vlastností austenitické ventilové oceli. Diplomová práce KMT – 116, Technická univerzita v Liberci.

[11] Jakubík, J. Změny mechanických vlastností austenitické ventilové oceli. Diplomová práce KMT – 114, Technická univerzita v Liberci.

PŘÍLOHA:

| | | |
|---|---|---|
| TEDOM | OPERAČNÍ NÁVODKA KONTROLY PRO PRACOVIŠTĚ **AJ231N** | Listů 2 |

| | | |
|---|---|---|
| Název dílu: ventil výfukový | Číslo výkresu: 1209 014, 027, 028 | Provoz: divize MOTORY<br>Jablonec n. N. 466 05 |
| Váha dílu: 0,27 kg | Druh materiálu: **1.4871** | Počet kusů: 6 ks na motor |

| | |
|---|---|
| Etalon a místa měření sondou:<br> | Požadavky na kontrolu:<br><br>Diagnostika dlouhodobé provozní teplotní degradace (zkřehnutí) austenitické oceli. Provádí se po demontáži hlav válců na čelech ventilů ještě v hlavách, nebo na demontovaných samostatných ventilech. |

| Název | Operace č. | Popis práce – sled úkonů | Zařízení, přípravky předmět | celkový čas s. | Displej |
|---|---|---|---|---|---|
| Nastavení | 1 | Připojit sondu k zařízení a zajistit šroubením.<br>Hlavní přepínač do polohy „ZAPNUTO“.<br>Přepínač magnetizace M do polohy 4.<br>Přepínač funkcí do polohy T | DOMENA B, NC | 15 | 000<br><br>1.00 |

| | | | | | |
|---|---|---|---|---|---|
| Začátek kalibrace | 2 | Klávesnicí napsat na displeji hodnotu etalonu 180 (Mo).<br>Zmáčknout tlačítko „ENTER" (šipka)<br>Dvěma prsty uchopit sondu při jejím čele a přiložit na předkreslené (kružnice) místo na povrchu etalonu. | Etalon Mo180 | 10 | 180<br>C 0 |
| Kalibrace dle etalonu | 3 | Zmáčknout dalším prstem tlačítko start na sondě, proběhne měřící cyklus, na displeji se ukáže hodnota vnitřní korekční konstanty × $10^{-3}$. | | 10 | C 1<br>C 2<br>C 3<br>Konst |
| Ověření kalibrace | 4 | Přepínač funkcí do polohy L<br>Zmáčknout start – proběhne měřící cyklus (neustále přidržovat sondu na etalonu)<br>Vypnutím přístroje se kalibrace zruší.<br>Displej musí ukazovat hodnotu Mo+-2%. Při větší odchylce nutno operace 2-4 opakovat. | | 5<br>Re 20 | 000<br>C 1<br>C 2<br>C 3<br>177-183 |
| Diagnostika oceli ventilu | 5 | Diagnostika ventilu se provádí ve čtyřech polohách po $90^0$ sondou čelem umístěnou mezi kalotou a obvodem čela (talíře) hlavy ventilů. Čelo zbavit pouze hrubých oddělitelných nánosů splodin.<br>Přepínač funkcí je v poloze měření L | Ventilová hlava | 4×5 =20 | |
| Měření M | 6 | Měřící čelo sondy musí být těsně přiloženo k talíři ventilu; zmáčknout „START". Velikost hodnoty M (intenzita A/m zbytkového magnetického pole na povrchu) je úměrná změnám vlastností oceli provozem. Nutno dohledat maximum M na čele. | | | 000<br>C 1<br>C 2<br>C 3<br>M |
| Vyhodnocení 1 ventilu | 7 | Nepoužitý ventil M max 10<br>Běžný provoz do 5000h (150 Mm) max 200 | | | |

| | | | | | |
|---|---|---|---|---|---|
| | | Dlouhodobý provoz silné zatížení M max 500<br>Ventil vyměnit – zkřehlý, lom SP. M nad 700<br>Při hodnotě M nad 1000 ukáže displej chybu | | | Er 0 |
| Vyhod-nocení ventilů motoru | 8 | Počet měření na všech ventilech 6x4<br>Vhodné provést 1 kontrolní měření i na sacích ven-tilech. Za běžného režimu jsou dostatečně chlazeny a hodnoty Mmax15. M nad 20 svědčí o špatném seřízení spalování. | | 150 | |
| celkem | | | | 210 | |

| | | |
|---|---|---|
| Vypracoval: Dne: SKRBEK. B. 2010-02-12. | Zkontroloval: Dne:<br>Skalský J. | Schválil: Dne: VRBA Vla-dimír 2010-03-02 |
| Změny | | |

# High temperature ultrasonic testing by Phased Array technique

Jana Veselá [2], Pavel Mares [14]
Research Centre Rez [7], Husinec-Rez, Czech Republic

**Abstract:** Laboratory ultrasonic testing of a welded joint was performer on an 89x6 mm pipe to verify the test system from point of view of the specific type of defects detection at the selected temperature. The test system includes ultrasonic, cooling and handling subsystems. The testing was performer in the temperature range from 50 °C to 350 °C in steps of 50 °C. The aim was to verify the high-temperature wedges, the cooling system, the high-temperature coupling means and the manip-ulator SKOOT HT with regards to the temperature, exposure time and space requirements of the entire system. The ultra-sonic system includes software designed to ultrasonic testing simulation up to a temperature of 350 °C. Based on the results, it is planned to carry out measurements at the nuclear power plant at operating temperatures that do not exceed 200 °C for the considered welded joints. As part of the testing, an inspection procedure and a safe working procedure were created, necessary for performing testing at high temperatures not only from the point of view of safety work, but also continuity and holding individual testing steps to maintain measurement repeatability.

# AUTHOR INDEX


Boat, Matthew ........ 111
Brackmann, Lukas ........ 58
Brom, Jaroslav ........ 1
Buethe, Inka ........ 58
Buschke, Paul ........ 102
Böhm, Stefan ........ 159
Cheng, L. ........ 22
Cheng, Liangliang ........ 88
De Fino, Mariella ........ 10
Deresse, Nuhamin E. ........ 29
Dilger, Klaus ........ 159
Fatiguso, Fabio ........ 10
Fiore, Alessandra ........ 10
Fousek, Petr ........ 153
Franzen, Andreas ........ 102
François, Stijn ........ 29
Grošek, Jiøí ........ 139
Hattori, Shinichi ........ 78
Hedayatrasa, Saeid ........ 88
Heilmann, Stephan ........ 39
Henrix, Frank ........ 102
Herzog, Thomas ........ 39
Heuer, Henning ........ 39
Ingale, T. ........ 68
Jankù, Michal ........ 139
Kamada, Toshiro ........ 78
Kenny, Stuart G. ........ 111
Kersemans, M. ........ 22
Kersemans, Mathias ........ 88
Kopec, B. ........ 123
Kryukov, Igor ........ 159
Mares, P. ........ 179
Mareš, Pavel ........ 145
Mazal, Pavel ........ 49
Meschke, Gunter ........ 58
Michauk, Jürgen ........ 39
Mohammad, Housam ........ 49
Mormile, Pasquale ........ 97
Mráz, Jakub ........ 171
Mueller, Inka ........ 58
Mund, Malte ........ 159
Muravev, Vitaly V. ........ 117
Muraveva, Olga V. ........ 117
Neugebauer, Josef ........ 153
Paegpegem, W. ........ 22
Patera, Jan ........ 1
Philipps, Uwe ........ 102
Priebe, Sebastian ........ 58
Prints, Eugen ........ 159
Rippa, Massimo ........ 97
Rottger, Arne ........ 58
Sarem, Mina ........ 29
Schubert, Frank ........ 39
Skrbek, Bøetislav ........ 171
Stryk, Josef ........ 139
Svoboda, Václav ........ 1
Terasawa, Koki ........ 78
Tong, Zongfei ........ 88
Van Steen, Charlotte ........ 29
Verstrynge, Els ........ 29
Vesela, J. ........ 179
Veselá, Jana ........ 145
Vlasic, Frantisek ........ 49
Walter, Susan ........ 39
Werner, Daniel ........ 102
Xie, Shejuan ........ 88
Yaghoubi, V. ........ 22
Yamashita, Kentaro ........ 78
Šrajbr, Christian ........ 159